Nanocatalysts

Edited by Indrajit Sinha and Madhulata Shukla

Published in London, United Kingdom

IntechOpen

Supporting open minds since 2005

Nanocatalysts
http://dx.doi.org/10.5772/intechopen.78514
Edited by Indrajit Sinha and Madhulata Shukla

Contributors
Sundaram Singh, Shivam Bajpai, Nobanathi Wendy Maxakato, Sandile Surprise Gwebu, Philiswa Nomngongo, Xiao-Lan Huang, Mohamad Fakhrul Ridhwan Samsudin, Nurfatien Bacho, Suriati Sufian, Tokeer Ahmad, Farha Naaz, Umar Farooq, Alex Tawanda Kuvarega, Bhekie Mamba, Myra Sarre Kadia Nzaba, Shivatharsiny Rasalingam, Sivagowry Shanmugaratnam, Indrajit Sinha, Alkadevi Verma, Madhulata Shukla

First published in London, United Kingdom, 2019 by IntechOpen
IntechOpen is the global imprint of INTECHOPEN LIMITED, registered in England and Wales, registration number: 11086078, The Shard, 25th floor, 32 London Bridge Street
London, SE19SG – United Kingdom
Printed in Croatia

British Library Cataloguing-in-Publication Data
A catalogue record for this book is available from the British Library

Additional hard and PDF copies can be obtained from orders@intechopen.com

Nanocatalysts
Edited by Indrajit Sinha and Madhulata Shukla
p. cm.
Print ISBN 978-1-78984-159-6
Online ISBN 978-1-78984-160-2
eBook (PDF) ISBN 978-1-78985-046-8

For EU product safety concerns:
IN TECH d.o.o., Prolaz Marije Krucifikse Kozulić 3, 51000 Rijeka, Croatia, info@intechopen.com or visit our website at intechopen.com.

Meet the editors

Dr. Indrajit Sinha is currently Associate Professor at the Department of Chemistry, Indian Institute of Technology (Banaras Hindu University, BHU), Varanasi. He obtained his Masters in Chemistry from IIT Delhi (India) and PhD from BHU. Areas of research include nanoparticles, catalysis, and computational physical chemistry. He has about 46 peer-reviewed research publications that include papers in reputed international journals, book chapters, etc.

Dr. Madhulata Shukla received her BSc and MSc degrees from Banaras Hindu University (BHU). She completed her PhD at the Department of Chemistry, Institute of Science, BHU, in 2014. After that she moved to the Department of Chemistry, IIT-BHU for National Post-Doctoral Research. At present she is a permanent faculty and head in the Department of Chemistry, G. B. College Ramgarh, Veer Kunwar Singh University, Bihar. Her present research interests are: synthesis, structure, and interaction in ionic liquids, spectroscopic study, density functional theory calculation, and nanoparticle synthesis in ionic liquids. She has 20 international publications in reputed journals and three book chapters. She is also the editorial board member of an international journal and life member of the Indian Science Congress Association and Chemical Research Society of India.

Contents

Preface

Nanocatalysis is a topical area of research that has huge potential. It attempts to merge the advantages of heterogeneous and homogeneous catalysis. The collection of articles in this book treats the topics of specificity, activity, reusability, and stability of the catalyst and presents a compilation of articles that focuses on different aspects of these issues. Thus, the first chapter of the book introduces and overviews the present status of understanding in the area of nanocatalysis. The tailoring of nanostructures to combine aspects of catalysis requires understanding of the mechanisms involved. For instance, the influence of size and shape of nanoparticles on their catalytic properties is specific to a particular reaction. Multimetallic nanoparticles provide an additional variety for inducing reaction-specific catalyst properties. Another chapter illustrates the successful use of zirconia nanoparticles as catalysts for the multicomponent reaction of isatin derivatives with ammonium acetate and aromatic aldehydes under solvent-free conditions. The authors emphasize the reusability and stability of these zirconia nanocatalysts for this class of reactions, and review the catalytic properties of functionalized and core–shell-type iron oxide-based magnetic nanoparticles. The chapter focuses on the effect of different types of functionalization on their catalyst properties for specific organic reactions. Platinum group-based nanoparticles are one of the most studied classes of catalysts applied to different types of reactions. This book includes a chapter emphasizing the use of such catalysts to facilitate reactions needed for environmental remediation. Another chapter talks about the nanomaterials that have been successfully used as catalysts for the preparation of compounds needed in composite solid propellants. Given the diversity of nanostructures and their applications treated here, we believe that this book will be an active source of information for research in the field of nanocatalysis.

Editor
Dr. Indrajit Sinha
Department of Chemistry,
Indian Institute of Technology (BHU),
Varanasi, India

Co-editor
Dr. Madhulata Shukla
Department of Chemistry,
G. B. College, Ramgarh,
Veer Kunwar Singh University, India

Chapter 1

Introductory Chapter: Salient Features of Nanocatalysis

Alkadevi Verma, Madhulata Shukla and Indrajit Sinha

1. Introduction

Drawbacks in homogeneous and heterogeneous catalysts necessitate new catalytic paradigms for overcoming the limitations associated with both types. The model catalyst should combine the advantages of homogeneous with heterogeneous catalysis. Thus, the catalyst for a particular reaction should exhibit good activity, selectivity, and product yield. At the same time, it should be separable (recoverable) from the reaction medium, stable and reusable. Tailored nanostructures have displayed the potential to meet these stringent requirements.

Transition metals, specifically precious noble metals such as Pt, Pd, Rh, Ru, Au, Ag, and Cu, are commonly used as homogeneous and heterogeneous catalysts in majority of chemical transformations [1, 2]. The primary reason for this is the variable oxidation states offered by them. They also possess good adsorption properties essential for heterogeneous catalysis. Combination of these two properties enables the transition metal nanoparticles to act as electron conduits for the reactants adsorbed on the surface of the catalyst. Initial examples of nanoparticles in catalysis were Ag nanoparticles in photography and Pt utilized in the decomposition of hydrogen peroxide (H_2O_2). Thereafter, noble metal nanoparticles have been used extensively as catalysts for many organic reactions such as carbon-carbon coupling in Suzuki, Stille and Heck reactions, hydrogenation, dehydrogenation reaction, oxidation, etc. [3].

Nanoparticles, owing to high surface energies, tend to get agglomerated resulting in enhanced particle sizes with lower surface area. The latter implies lesser number of surface active sites in the catalyst. Stabilizers such as surfactants or polymers, that may also act functionalizing agents, are frequently used to protect nanoparticles surfaces against aggregation. Such surface altering processes also cause change in the electronic structure of the nanoparticle and because of that in their catalytic activity as well [4]. The other approach to circumvent this problem is by implanting these NPs on large surface area but low density insoluble solids supports like zeolites, carbon based materials etc. The support material may be relatively inert. Alternatively, the support could modify the chemical and adsorption properties of the catalyst. Active supports like these may enhance of impede the performance of the catalyst for a specific reaction by tuning the electron density of NPs. Another possible scenario is that the support is a better adsorbent for one of the reactants and thereby improves the performance of the catalyst.

Currently, nanoparticles are increasingly substituting conventional heterogeneous catalysts [5]. Due to smaller sizes, nanoparticles have higher surface area and increased exposed active sites. In that way nanoparticles have larger contact areas with reactants and are catalytically more active than conventional heterogeneous catalysts. Variations in shape and composition of nanocatalysts give access to

IntechOpen

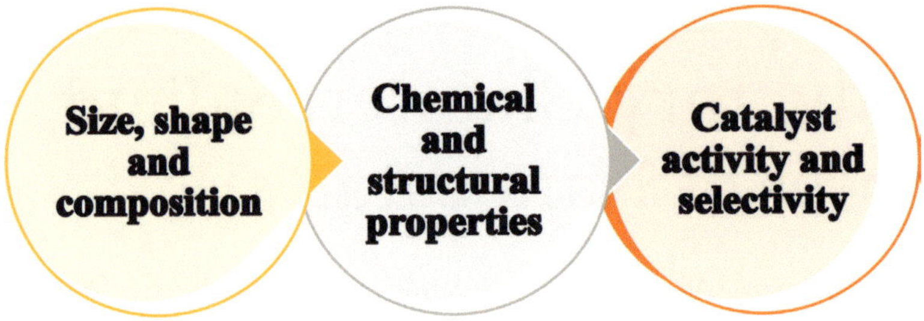

Figure 1.
Dependence of catalytic activity on size, shape and composition.

different types of catalytic sites. A particular type of site displays better selectivity towards a particular reaction pathway. Thus, from the point of view of increased activity and selectivity nanocatalysts have properties which tend to those of homogeneous catalysts. On the other hand, nanocatalysts are relatively easier to separate from the reaction mixtures and therefore, in that sense, are heterogeneous catalysts. Furthermore, adsorption of reactant(s) on to the nanocatalyst is a necessary precondition for any nanocatalyzed reaction. This is again characteristic of a heterogeneous catalytic process. Therefore, nanocatalysts with better activity, stability, and selectivity can be designed and synthesized by controlling their size, shape, and composition of nanomaterials [6–8]. **Figure 1** illustrates the typical cause and effect relation followed.

To study the size effect of catalyst, metal nanoparticles with the same shape but different sizes are applied in a reaction. The influence of nanoparticle size on catalytic activity and selectivity can thus be determined.

2. Effect of size

2.1 On catalytic properties

Nanocatalysts as compared to their bulk counterparts, commonly offers much higher surface-to-volume ratio. Prominent changes in the electronic states and coordination environment of the surface atoms of a catalyst nanoparticle might be possible when its size decreases typically to a certain nanoregime. Therefore, change in size of nanoparticles affects coordination environment, electronic state, and adsorption energy of the reactant molecules.

2.2 Size-dependent coordination environment

The effect of atoms at corners and edges of nanoparticles becomes dominant with decreasing the size of nanoparticles [9, 10]. Cao et al. summarized a relation between surface metal atoms with different coordination numbers of cuboctahedral and cubic geometry of nanoparticles with overall size of the nanoparticles [11]. They concluded that the coordination numbers 9, 7, and 4 of a cuboctahedral nanoparticle and 8, 6, and 3 in a cubic nanoparticle exhibits strong dependence on the size of the nanoparticle. Such strong correlation of size-dependent catalytic performance (for a particular nanocatalyst shape) was also reported by Tao et al. for room temperature CO oxidation reaction. For instance, in Pt nanoparticles with a size of about 2.2 nm, the Pt atoms (CN = 7) at the edge of triangular nanoclusters are active for CO oxidation even at room temperature. However, Pt atoms with CN of 9 on the terrace of Pt (111) are not active for CO oxidation at room temperature [12].

2.3 Size-dependent electronic state

The electronic structure of metal nanoparticles of 1–2 nm (in the quantum regime) is like that of a molecule. Thus, Au nanoparticles smaller than 1 nm, are more molecular than metallic. Thus, molecule-like electronic states of metal nanoparticles of 1–2 nm exhibits inherently different catalytic performance in contrast to a nanoparticle with a larger size [11]. This was experimentally demonstrated for the first time by Goodman et al., in CO oxidation on Au nanocluster with thickness of three atomic layers supported on TiO_2 [13, 14]. Analysis of Au LIII XANES white lines by these authors revealed that supported Au nanoparticles with different sizes have different average coordination numbers. Thus Au nanoparticle of 3 nm has average CN = 9.5. Similarly the nanoparticles of 1 nm have average CN = 6, while nanoparticles of 0.5–1 nm have CN = 3.6. This shows that smaller Au nanoparticles have a size-dependent electronic environment [15–17].

2.4 Size-dependent adsorption energy

Adsorption is a primary step in heterogeneous catalysis. Size-dependent adsorption energies of reactants on catalyst surfaces with different coordination numbers have also been suggested in literature. References [18, 19] assert that the adsorption energy is dependent on the coordination environment of metal nanoparticles. Usually, catalyst atom(s) with low coordination number (CN) exhibits stronger adsorption for a given molecule than those catalyst atoms with higher coordination number [20, 21]. For example, adsorption energy of adsorbates including $^{\bullet}O_2$, $^{\bullet}OH$, $^{\bullet}OOH$, $^{\bullet}O$, $^{\bullet}H_2O$, and $^{\bullet}H_2O_2$ on Pt nanocatalyst decrease linearly with increase in coordination number from 3 to 9 [20]. Similar linear relationships between adsorption energy and coordination number have been reported for other transition metal catalysts such as Co, Ni, Cu, Rh, Pd, Ag, Ir, and Au [20].

3. Effect of shape on catalytic properties

The representative shapes of metal nanoparticles based on dimensionality are shown in **Figure 2**. Spherical, pseudo-spherical, dodecahedral, tetrahedral, octahedral, cubic shape represents 0D nanoparticles. 1D morphology of nanoparticles includes nanotubes, nanorods or nanowires, nanocapsules, etc. [21, 22]. Hexagonal, triangular, quadrangular plates or sheets, belts, rings, etc. fit in to the 2D shape NPs [23]. 3D morphologies of nanoparticles are complex such as nanoflowers, nanostars, polygonal nanoframes, etc. [24, 25]. Compared to simple

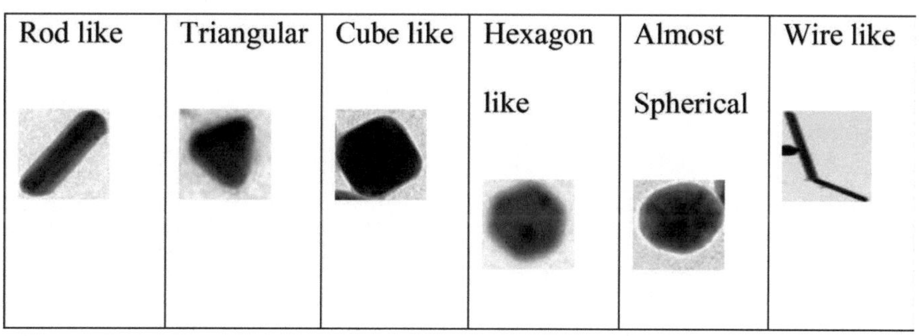

Figure 2.
Different types of anisotropic nanoparticles.

isotropic morphologies of nanoparticles, novel anisotropic morphologies have unique physicochemical properties due to the different numbers of steps, edges, and kink sites present on to the surface of catalyst in nanoscale regime. For example, polyhedral Au NPs with high-indexed facets are found to exhibit excellent optical and catalytic properties, [26, 27] Au rods with different ratios of length and width display different transverse and longitudinal plasmon bands. Preicel et al. have recently published a review on different types of anisotropic gold nanoparticles used in catalysis [28]. Branched Au NPs with multiple tips adopting structures like stars and flowers are increasingly being used for catalysis, surface-enhanced Raman scattering, and sensing [29].

4. Composition effect

The section introduces the effect of composition on catalytic activity from the perspective of alloy and bimetallic nanoparticles only. Commonly, bimetallic nanoparticles can be categorized into alloy (ordered or random), Janus and core-shell (core shell or cluster-in-cluster) structure types. The type of bimetallic or alloy nanostructure formed depends on the synthesis methodology utilized (**Figure 3**).

Catalytic activity of bimetallic nanomaterials is different from of its component metals. Instead of being an average of the catalytic activities of its components, bimetallic nanoparticles may also exhibit synergistic catalytic properties [30, 31]. One such example of composition effect was studied by Lim and co-workers in catalytic activity of Pt-Y alloy for electrocatalytic oxygen reduction [32]. The addition of various amounts of Y changes the electronic structure of Pt and thus modifies the binding energy of the oxygen-containing species. The optimum catalytic performance was achieved at a particular composition of Pt-Y alloy. Thus, the catalytic activity of Pt-Y alloy catalysts follows the trend of $Pt_{70}Y_{30} > Pt_{78}Y_{22} > Pt_{64}Y_{36} > Pt_{86}Y_{14} > Pt_{91}Y_{9} > Pt$. Sun and co-workers also demonstrated such composition-dependent catalytic activity of monodisperse CoPd nanoparticles for formic acid oxidation [33].

The effect of composition also exists in core-shell bimetallic nanoparticles. Jiang et al. established the composition-dependent activity of core-shell Cu@M (M = Co, Fe, Ni) catalyst nanoparticles for hydrolytic dehydrogenation of ammonia borane [34]. In core-shell Cu@M structures, collaboration of Cu with M can change the width of surface d band, which is beneficial for catalytic enhancement. Only an optimum Cu/M ratio in all three cases shows the best catalytic activity.

Extensive use of BNPs have been reported in catalytic oxidation of dyes [35], glucose [29], CO [36], benzyl alcohol [37], and methanol [38] oxygen reduction [39] propane dehydrogenation [36] hydrogenation of nitro-aromatic compounds [40], electro-catalytic oxidation of methanol [28] as well as in desulfurization of thiophene [41].

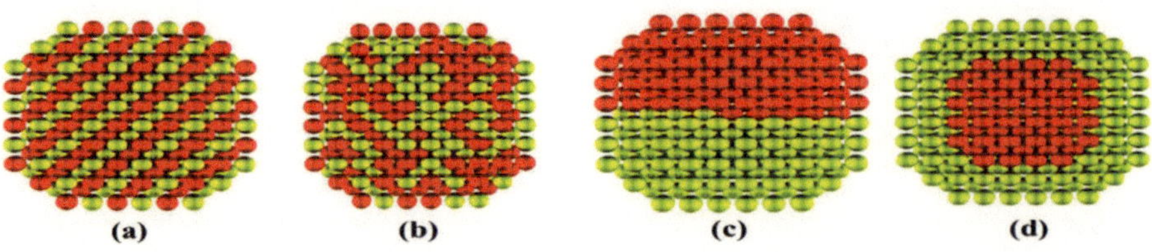

Figure 3.
Different possibilities of bimetallic nanostructures observed: (a) ordered alloy; (b) random alloy; (c) Janus-like; and (d) core shell.

DOI: http://dx.doi.org/10.5772/intechopen.86209

5. Conclusions

High surface area and consequently, enhanced surface active sites have led to extensive use of nanoparticles (NPs) as catalysts. Altering the nature and density of active sites can improve their catalytic activity. Change in nanoparticle size, shape, and composition affects the active site catalytic properties. The three listed aspects may also affect the electronic structure of the nanostructures. Moreover, appropriate functionalization of the nanostructures, not improve their stability against aggregation, but also impact their electronic structures and adsorption properties. Low density support materials also influence the nanostructure electronic state and the resultant properties.

Anisotropic shapes offer different densities of surface, edge, and corners in nanoparticles. Atoms in corners and edges possess low coordination and can lead to better interaction with the substrate and other reacting species for catalysis. On the other hand, possible variations in composition offered by bimetallic nanoparticles can not only reduce the cost of nanomaterial but may also show synergistic properties.

Author details

Alkadevi Verma[1], Madhulata Shukla[2] and Indrajit Sinha[3*]

1 Department of Chemistry, Rewa Engineering College, Rewa, India

2 Department of Chemistry, G.B. College, Veer Kunwar Singh University, Ramgarh, India

3 Department of Chemistry, Indian Institute of Technology (Banaras Hindu University) Varanasi, India

*Address all correspondence to: isinha.apc@iitbhu.ac.in

References

[1] Haruta M. Catalysis of gold nanoparticles deposited on metal oxides. CATTECH. 2002;**6**(3):102-115

[2] Bayo'n JC, Claver C, Masdeu-Bulto AM. Homogeneous catalysis with transition metal complexes containing sulfur ligands. Coordination Chemistry Reviews. 1999;**193-195**:73-145

[3] Hutchings G. Nanocatalysis: Synthesis and applications. Hoboken, New Jersey, USA: John Wiley & Sons. Inc; 2013

[4] Shukla M, Sinha I. Catalytic activation of nitrobenzene on PVP passivated silver cluster: A DFT investigation. International Journal of Quantum Chemistry. 2018;**118**:e25490

[5] Ozin GA, Arsenault AC, Cademartiri L. Nanochemistry: A Chemical Approach to Nanomaterials. London, United Kingdom: Royal Society of Chemistry; 2009

[6] Astruc D, editor. Nanoparticles and catalysis. Hoboken, New Jersey, USA: John Wiley & Sons; 2008

[7] Ishida T, Haruta M. Gold catalysts: Towards sustainable chemistry. Angewandte Chemie International Edition. 2007;**46**(38):7154-7156

[8] Hashmi ASK, Hutchings GJ. Gold catalysis. Angewandte Chemie International Edition. 2006;**45**(47):7896-7936

[9] Hvolbæk B, Janssens TV, Clausen BS, Falsig H, Christensen CH, Nørskov JK. Catalytic activity of Au nanoparticles. Nano Today. 2007;**2**(4):14-18

[10] Che M, Bennett CO. The influence of particle size on the catalytic properties of supported metals. Advances in Catalysis. 1989;**36**:55-172

[11] Cao S, Tao FF, Tang Y, Li Y, Yu J. Size and shape-dependent catalytic performances of oxidation and reduction reactions on nanocatalysts. Chemical Society Reviews. 2016;**45**(17):4747-4765

[12] Tao F, Dag S, Wang LW, Liu Z, Butcher DR, Bluhm H, et al. Break-up of stepped platinum catalyst surfaces by high CO coverage. Science. 2010;**327**(5967):850-853

[13] Chen MS, Goodman DW. The structure of catalytically active gold on titania. Science. 2004;**306**(5694):252-255

[14] Valden M, Lai X, Goodman DW. Onset of catalytic activity of gold clusters on titania with the appearance of nonmetallic properties. Science. 1998;**281**(5383):1647-1650

[15] Van Bokhoven JA, Miller JT. Electron density and reactivity of the d band as a function of particle size in supported gold catalysts. The Journal of Physical Chemistry C. 2007;**111**(26):9245-9249

[16] Janssens TV, Carlsson A, Puig-Molina A, Clausen BS. Relation between nanoscale Au particle structure and activity for CO oxidation on supported gold catalysts. Journal of Catalysis. 2006;**240**(2):108-113

[17] Miller JT, Kropf AJ, Zha Y, Regalbuto JR, Delannoy L, Louis C, et al. The effect of gold particle size on Au-Au bond length and reactivity toward oxygen in supported catalysts. Journal of Catalysis. 2006;**240**(2):222-234

[18] Somorjai GA, Li Y. Introduction to Surface Chemistry and Catalysis. Hoboken, New Jersey, USA: John Wiley & Sons; 2010

[19] Somorjai GA. Chemistry in Two Dimensions: Surfaces. Ithaca, NY, USA: Cornell University Press; 1981

[20] Calle-Vallejo F, Loffreda D, Koper MT, Sautet P. Introducing structural sensitivity into adsorption–energy scaling relations by means of coordination numbers. Nature Chemistry. 2015;7(5):403-410

[21] Lu Z, Huang Y, Zhang L, Xia K, Deng Y, He N. Preparation of gold nanorods using 1,2,4-Trihydroxybenzene as a reducing agent. Journal of Nanoscience and Nanotechnology. 2015;**15**(8):6230-6235

[22] Zhang L, Xia K, Lu Z, Li G, Chen J, Deng Y, et al. Efficient and facile synthesis of gold nanorods with finely tunable plasmonic peaks from visible to near-IR range. Chemistry of Materials. 2014;**26**(5):1794-1798

[23] Li J, Liu CY, Liu YA. Graphene hydrogel: Synthesis, characterization and its use for catalytic reduction of 4-nitrophenol. Journal of Materials Chemistry. 2012;**22**(17):8426-8430

[24] Lv H, Ji G, Liu W, Zhang H, Du Y. Achieving hierarchical hollow carbon@Fe@Fe_3O_4 nanospheres with superior microwave absorption properties and lightweight features. Journal of Materials Chemistry C. 2015;**3**(39):10232-10241

[25] Lin X, Ji G, Liu Y, Huang Q, Yang Z, Du Y. Formation mechanism and magnetic properties of hollow Fe_3O_4 nanospheres synthesized without any surfactant. CrystEng Comm. 2012;**14**(24):8658-8663

[26] Tran TT, Lu X. Synergistic effect of Ag and Pd ions on shape-selective growth of polyhedral Au nanocrystals with high-index facets. The Journal of Physical Chemistry C. 2011;**115**(9):3638-3645

[27] Jing H, Zhang Q, Large N, Yu C, Blom DA, Nordlander P, et al. Tunable plasmonic nanoparticles with catalytically active high-index facets. Nano Letters. 2014;**14**(6):3674-3682

[28] Priecel P, Salami HA, Padilla RH, Zhong Z, Lopez-Sanchez JA. Anisotropic gold nanoparticles: Preparation and applications in catalysis. Chinese Journal of Catalysis. 2016;**37**(10):1619-1650

[29] Zhang H, Toshima N. Synthesis of Au/Pt bimetallic nanoparticles with a Pt-rich shell and their high catalytic activities for aerobic glucose oxidation. Journal of Colloid and Interface Science. 2013;**394**:166-176

[30] Zhang H, Okuni J, Toshima N. One-pot synthesis of Ag–Au bimetallic nanoparticles with Au shell and their high catalytic activity for aerobic glucose oxidation. Journal of Colloid and Interface Science. 2011a;**354**(1):131-138

[31] Yin AX, Min XQ, Zhang YW, Yan CH. Shape-selective synthesis and facet-dependent enhanced electrocatalytic activity and durability of monodisperse sub-10 nm Pt-Pd tetrahedrons and cubes. Journal of the American Chemical Society. 2011;**133**(11):3816-3819

[32] Yoo SJ, Kim SK, Jeon TY, Hwang SJ, Lee JG, Lee SC, et al. Enhanced stability and activity of Pt-Y alloy catalysts for electrocatalytic oxygen reduction. Chemical Communications. 2011;**47**(41):11414-11416

[33] Mazumder V, Chi M, Mankin MN, Liu Y, Metin O, Sun D, et al. A facile synthesis of MPd (M = Co, Cu) nanoparticles and their catalysis for formic acid oxidation. Nano Letters. 2012;**12**(2):1102-1106

[34] Jiang HL, Akita T, Xu Q. A one-pot protocol for synthesis of non-noble

metal-based core-shell nanoparticles under ambient conditions: Toward highly active and cost-effective catalysts for hydrolytic dehydrogenation of NH_3 BH_3. Chemical Communications. 2011;**47**(39):10999-11001

[35] Jiang HL, Akita T, Ishida T, Haruta M, Xu Q. Synergistic catalysis of Au@ Ag core-shell nanoparticles stabilized on metal-organic framework. Journal of the American Chemical Society. 2011;**133**(5):1304-1306

[36] Xia Y, Xiong Y, Lim B, Skrabalak SE. Shape-controlled synthesis of metal nanocrystals: Simple chemistry meets complex physics? Angewandte Chemie International Edition. 2009;**48**(1):60-103

[37] Han Z, Li S, Jiang F, Wang T, Ma X, Gong J. Propane dehydrogenation over Pt–Cu bimetallic catalysts: The nature of coke deposition and the role of copper. Nanoscale. 2014;**6**(17):10000-10008

[38] Deplanche K, Merroun ML, Casadesus M, Tran DT, Mikheenko IP, Bennett JA, et al. Microbial synthesis of core/shell gold/palladium nanoparticles for applications in green chemistry. Journal of the Royal Society Interface. 2012;**9**(72):1705-1712

[39] Fu GT, Ma RG, Gao XQ, Chen Y, Tang YW, Lu TH, et al. Hydrothermal synthesis of Pt–Ag alloy nano-octahedra and their enhanced electrocatalytic activity for the methanol oxidation reaction. Nanoscale. 2014;**6**(21):12310-12314

[40] Yamamoto TA, Kageyama S, Seino S, Nitani H, Nakagawa T, Horioka R, et al. Methanol oxidation catalysis and substructure of PtRu/C bimetallic nanoparticles synthesized by a radiolytic process. Applied Catalysis A: General. 2011;**396**(1):68-75

[41] Suo Z, Ma C, Liao W, Jin M, Lv H. Structure and activity of Au–Pd/ SiO_2 bimetallic catalyst for thiophene hydride sulfurization. Fuel Processing Technology. 2011;**92**(8):1549-1553

Chapter 2

Platinum-Based Carbon Nanodots Nanocatalysts for Direct Alcohol Fuel Cells

Sandile Surprise Gwebu, Philiswa N. Nomngongo and Nobanathi W. Maxakato

Abstract

Platinum and its alloys are regarded as best nanocatalysts for the electro-oxidation of alcohol fuels especially in acidic conditions. The performance of nanocatalysts for alcohol fuel cells depends greatly on the support material. A good support material should have high surface area to obtain high metal dispersion. It should also bond and interact with the nanocatalysts to improve the activity of the overall electrode. Most importantly, the support material should offer great resistance to corrosion under the harsh fuel cell conditions. In this chapter, the use of carbon nanodots as support materials for Pt-Sn and Pt-TiO_2 nanoparticles is discussed. The electrochemical activity of Pt/CNDs, Pt-Sn/CNDs and Pt/CNDs-TiO_2 nanocatalysts was studied using cyclic voltammetry (CV) in acidic and alkaline conditions. Chronoamperometry (CA) was used to investigate the long-term stability of the nanocatalysts under the fuel cell environment. Electrochemical results demonstrated that binary Pt nanocatalysts are more active compared to monocatalysts. It was also observed that carbon nanodots are better support materials than carbon black. Blending carbon nanodots with titanium dioxide (a ceramic material) improves the corrosion resistance of the nanocatalyst. Cyclic voltammetry results also proved that alcohol electro-oxidation is enhanced in alkaline conditions.

Keywords: nanocatalyst, platinum, carbon nanodots, alcohol fuel cell, electrooxidation

1. Introduction

A fuel cell is an electrochemical device that transforms the chemical energy from a fuel into electrical energy through an electrochemical reaction without combusting the fuel [1, 2]. Fuel cell technology is one of the innovative future energy technologies all over the world. Fuel cells are environmentally friendly and more efficient than combustion engines. They operate at low temperatures quietly without vibration or noise; hence, can be used for both stationary and mobile applications. There are numerous types of fuel cells, classified according to the type of the electrolyte they use which governs the operating temperature and the catalyst suitable for those conditions.

Among the numerous types of fuel cells, direct alcohol fuel cells (DAFCs) are the most attractive using liquid and renewable alcohols (ethanol and methanol) as a fuel [3]. DAFCs are a very promising type when fuel consumption and feed strategies are taken into consideration. Unlike hydrogen fuel cells, which have safety, storage, and distribution challenges, DAFCs employ a liquid fuel that can easily be kept and transported [4, 5]. The electro-oxidation of alcohol fuels occurs on the surface of the positive electrode (anode) whereas the reduction of oxygen takes place on the negative electrode (cathode). DAFCs electrodes are made of conductive support material with a high surface onto which platinum nanoparticles are deposited. Platinum is the best electrocatalyst for alcohol oxidation reactions (AORs) in acidic conditions, nonetheless, it is expensive and it can effortlessly be poisoned by the intermediates that are formed during the AORs [6]. For these reasons, novel catalyst materials and catalyst preparation methods are still key topics studied in alcohol fuel cells. The efficiency of the electrocatalyst nanoparticles depends on the support material and its morphology [7]. Carbon nanomaterials have been considered for use as support materials because of their large surface area, great electrical conductivity, tensile strength, good thermal stability, corrosion resistance, ease of recovery and recyclability [8, 9]. In this chapter, the electrochemical performance of Pt/CNDs, Pt-Sn/CNDs and Pt/CNDs-TiO_2 nanocatalysts is discussed. All the metal nanoparticles used in the studies reviewed were deposited on to the support materials by the alcohol reduction method.

2. Direct alcohol fuel cells (DAFCs)

DAFCs belong to either polymer electrolyte membrane (PEM) or alkaline fuel cells (AFCs) depending on the pH of the electrolyte. Low molecular weight alcohol fuels are more attractive compared to pure hydrogen because they can be handled and transported easily. The design of DAFCs is greatly simplified; no reformer or humidifier is required. **Figure 1** shows a simple illustration of DAFC components.

Electrochemical reactions take place in the catalyst layers, which are attached to the sides of the membrane. The catalyst layers are designed in a way that can facilitate the movement of protons, electrons and reactants. The membrane serves

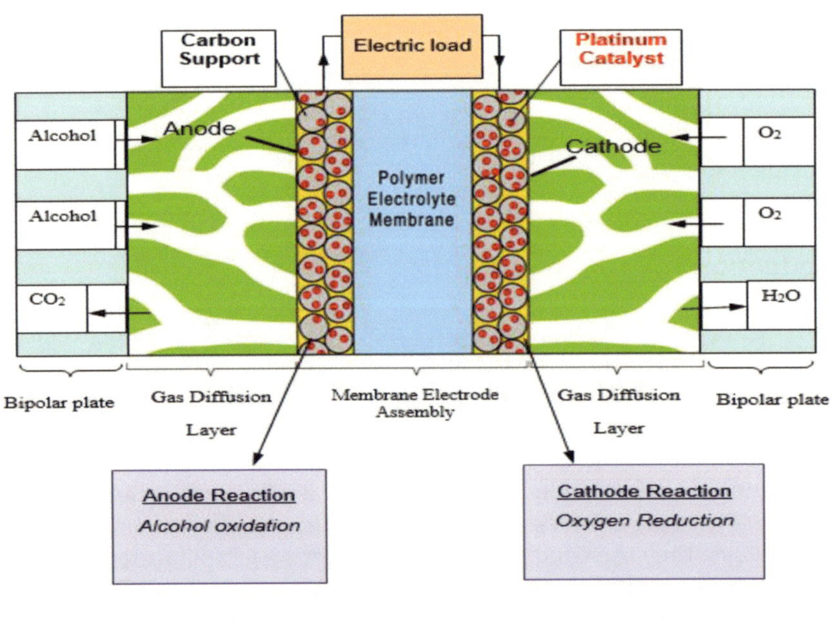

Figure 1.
Illustration of direct alcohol fuel cell components.

as an insulator for electrons and promotes the movement of protons from the anode to the cathode. The alcohol fuel is fed into the anode compartment where it gets oxidised, at the cathode side oxygen is reduced, both reactions produce electrons that migrate along the circuit and serves an electric load. The protons and other ions combine to form by-products, water and carbon dioxide.

The commercialisation of the DAFC technology has been hindered by (i) the inflated cost of the system, (ii) slow electro-oxidation reaction kinetics and fuel crossover. Platinum is the best electrocatalyst particularly in acid media, but it is very expensive. It is required in both electrodes of the fuel cell. Methanol is oxidised to produce CO, which poisons the platinum electrocatalyst. The poisoning of the platinum catalyst surface by carbon monoxide results in sluggish electro-kinetics of methanol electro-oxidation and this increases the rate of methanol crossover through the membrane. Carbon monoxide is the most stable intermediate formed during methanol electro-oxidation. It strongly adsorbs on the surface of the electrocatalyst, therefore preventing further methanol electro-oxidation. The removal of CO is the slowest reaction, hence, becomes the rate-limiting step. Ethanol is incompletely oxidised in acidic media producing intermediates such as CH_3CHO and CH_3COOH liberating only 2 and 4 electrons, respectively, in alkaline conditions it is selectively oxidised to acetate, this greatly reduces the overall fuel cell performance. Fuel crossover is a challenge, especially in acidic conditions. The simultaneous electro-oxidation of fuel and the reduction of oxygen at the cathode results in mixed potential and reduces the cell voltage. The fuel crossover also results in fuel wastage. In addition to this, the platinum surface gets poisoned by intermediates produced from alcohol electro-oxidation. All these factors diminish the overall fuel cell output.

2.1 Direct alcohol fuel cell working principle

At the anode side, alcohol fuel (methanol, ethanol) is oxidised with the aid of the electrocatalyst, generating electrons (e^-) and protons (H^+) according to the following equations:

In acidic conditions

$$CH_3OH + H_2O \rightarrow 6H^+ + 6e^- + CO_2 \quad (1)$$

$$C_2H_5OH + 3H_2O \rightarrow 2CO_2 + 12H^+ + 12e^- \quad (2)$$

In alkaline conditions

$$CH_3OH + 8OH^- \rightarrow CO_3^{2-} + 6e^- + 6H_2O \quad (3)$$

$$C_2H_5OH + 12OH^- \rightarrow 2CO_2 + 12e^- + 9H_2O \quad (4)$$

Electrons migrate to the cathode side through the external circuit, resulting in an electrical current. The protons diffuse to the cathode through the electrolyte and oxygen is reduced by these electrons to form anions, which react with the protons (H^+) to form water according to the following equations:

In acidic conditions

$$3O_2 + 12H^+ + 12e^- \rightarrow 6H_2O \quad (5)$$

In alkaline conditions

$$O_2 + 4e^- + 2H_2O \rightarrow 4OH^- \quad (6)$$

2.2 Support materials for fuel cell nanocatalysts

Support materials play a vital role in fuel cells as they strongly impact on the performance, stability and effectiveness of the electrocatalyst. Support materials can be classified as either primary or secondary. Primary supports are novel nanostructured carbon based such carbon nanotubes and carbon nanodots whereas secondary supports are metal oxides, which have been used mainly to modify, promote the primary supports and improve corrosion resistance. Corrosion of carbon support material is one of the major problems hindering the commercialization of direct alcohol fuel cells. A good support material must exhibit the following properties:

- High surface area to obtain maximum and uniform metal dispersion.
- Satisfactory electrical conductivity so that it can serve as a path for the movement of electrons.
- Suitable porosity to boost electrolyte flow.
- Electrochemical stability in the fuel cell environment.
- Availability and low cost.
- Suitable functional groups to enable bonding and interaction with the nanocatalysts to increase the activity and durability of the metal nanoparticles.

2.2.1 Carbon supports

Carbon is unique among all elements in the periodic table. It exists in radically different forms with variable micro-textures. This makes carbon an attractive material suitable for wide ranges of electrochemical applications [9]. Carbon is usually used for supporting nano-sized electrocatalyst particles for fuel cells owing to its excellent electrical conductivity, huge surface area, pore diameter and pore volume [10]. Carbon exists in many allotropic forms owing to its valency of 4 electrons. The most common types of carbon are carbon black, fullerenes, diamond, graphene and carbon nanotubes. Carbon black, graphene and carbon nanotubes have been extensively used as supports for fuel cell nanocatalysts. Carbon black is amorphous, usually contains spherical shaped graphite particles, characteristically with sizes below 50 nm. These particles combine to form aggregates and agglomerates with diameters around 250 nm [8]. Carbon black is mainly manufactured by the "furnace black" process, this process involves fractional oxidation of petrochemical or coal tar oils [11]. There are many varieties of carbon blacks, the most common types are vulcan XC-72, ketjen black and acetylene black. These carbon blacks exhibit good physicochemical properties such as surface functional groups, electrical conductivity, porosity and a reasonable amount of surface area. Carbon black is commonly used as electrocatalyst support material in fuel cells but still suffers from stability issues that reduce the electro-activity of the electrode [12, 13]. Carbon black is very unstable in highly acidic/alkaline media of a fuel cell, it gets corroded resulting in agglomeration and detachment of platinum catalyst, this decreases the overall fuel cell performance [14, 15].

Carbon nanotubes (CNTs) are cylindrical and hollow graphite layers with closed ends. They exist as either single wall or multi-walled nanotubes, with thicknesses of a few nanometres and lengths of about 1 mm [8]. A fascinating property of carbon nanotubes is their capability to capture other elements/nanocatalysts inside their wall structure. CNTs (single or multi-walled) are also used as catalysts supports in low-temperature fuel cells [16–19]. SWCNTs have huge surface areas whereas MWCNTs are better electrical conductors than SWCNTs [20]. CNTs have chemically unreactive surfaces making it difficult to deposit metal nanoparticles on their surfaces, hence, metal nanoparticles on CNTs generally have poor dispersion and large particle sizes, this decreases the electro-activity of the nanocatalyst [21]. Researchers have developed functionalisation methods in order to incorporate oxygen-containing groups on CNTs so as to improve their reactivity and interaction with nanocatalysts [15, 40]. These functionalisation methods include impregnation [22], acid treatment (such as a mixture of HNO_3 and H_2SO_4) [23, 24], polyol processing [25], ion-exchange [26] and electrochemical deposition [27].

Carbon nanofibers (CNFs) are industrially produced by the decomposition and graphitization of rich organic carbon-containing polymers. CNFs are also used as catalysts supports in fuel cells [28–30]. They have a thin cavity and some CNFs have no hollow cavity. CNFs are much thicker than CNTs. They can be classified into six types according to their structures which are; ribbon-like, platelet, herringbone, tubular, faceted tubular and faceted tubular multiwall [31]. Herringbone CNFs exhibit intermediate features of parallel and platelet types, thus providing higher catalytic activity and better durability than the other types [32].

The major difference between CNFs and CNTs is the exposure of reactive edge planes. Anchoring sites for the nanocatalysts resulting from the edge planes are exposed in CNFs while the basal planes are exposed in CNTs. Recent studies reveal that platinum electrocatalyst supported on CNFs show lower poisoning rates than platinum electrocatalyst supported on carbon black [31].

2.2.2 Carbon nanodots

Carbon nanodots (CNDs) are a newest type in the carbon nano family with particle diameters less than 10 nm. They were first obtained during the purification of SWCNTs through preparative electrophoresis in 2004 [33]. CNDs have gradually attracted too much attention because of their availability and inexpensive nature. Carbon nanodots core is sp^2 conjugated and it contains multiple oxygen moieties such as the carboxyl (—COOH), hydroxyl (—OH), and aldehyde (—COH) groups [34]. Synthesis of carbon nanodots can be categorised into two groups namely; top down (chemical) and bottom up (physical) methods [35, 36]. Top-down method employs treating starting materials such as graphitic powder or MWCNTs in harsh chemical conditions [36]. Bottom up approaches include ultra-sonication [37], microwave pyrolysis [38] and hydrothermal treatment of small molecules such as starch [39], citric acid [40] glucose [37] and leeks [41]. Carbon nanodots used in the studies described in this chapter were prepared by the pyrolysis of oats grains [42]. Pyrolysis is a bottom up (physical) method.

In the same study we used the freshly prepared CNDs to deposit Pt nanoparticles. The prepared Pt/CNDs electrocatalyst outperformed the platinum nanoparticles supported on carbon black. Wei et al. prepared naked palladium nanoparticles supported on carbon nanodots for methanol electro oxidation. It was observed that freshly prepared CNDs act as reductants resulting in attachment of metal nanoparticles on their surfaces [37].

What makes carbon nanodots superior?

- They are easy to synthesise from various starting materials such as carbohydrates and waste carbonaceous sources such paper and avocado seed.

- No need to functionalise-they contain oxygen content of about 10% wt.
- Unlike carbon nanotubes, carbon nanodots are non-toxic and non-carcinogenic.
- They are biocompatible.
- Due to their small particle sizes (below 10 nm) they provide high surface area which enhances metal nanoparticle dispersion.
- Carbon nanodots also bond with the metal nanoparticles; thus, improving the interaction between the catalyst and the support material.

2.2.3 Inorganic support materials

Inorganic metal oxides such as TiO_2 [15, 43–47], WO_3 [48–50], CeO_2 [29, 30, 51], Al_2O_3 [52] SnO_2 [53] and MoO_3 [54] are very stable under the harsh fuel cell environment. These ceramic materials are not good conductors of electricity; hence, they can be only utilized as secondary support materials; they cannot be used on their own as nanocatalyst supports.

2.2.3.1 Titanium dioxide (TiO_2)

Titanium dioxide commonly referred to as "titania" is a naturally occurring oxide of titanium. Titanium oxides have been employed in numerous applications that depend on its photo-electrochemical, catalytic and outstanding corrosion resistance in both alkaline and acidic media. Titania generally exists in three major forms which are; anatase, rutile, and brookite crystallography. Each of these structures exhibits unique physical properties which make them suitable for different applications. It has been reported that the anatase crystallography is more effective as an electrocatalyst than rutile titania. The chemical and physical properties displayed by these crystallographic forms depend on the synthesis route [55]. Titanium dioxide nanoparticles can be produced by approaches such as the sol-gel technique [55, 56], microwave assisted hydrolysis [57], hydrothermal method [58], co-precipitation route [59], flame combustion method [60, 61] and chemical vapour deposition (CVD) [62].

3. Nanocatalysts for direct alcohol fuel cells

3.1 Platinum (Pt) electrocatalyst

Platinum monometallic catalyst has been widely used in fuel cells and has shown good results for both DMFC and DEFC in acidic media [63–69]. This is because platinum shows better activity, selectivity, stability and resistance to poison than other metals. Platinum catalyst allows chemical bonds to cleave but weakly enough to yield the product when the reaction has taken place, this is a required property for a good heterogeneous catalyst [70]. However, the crystalline orientation of platinum is a very important factor that governs the activity and poisoning rates of the electrode.

Platinum crystallises in face-centred cubic structure (FCC) and the shape of the resulting facet is predicted using Wulff's rule [71]. The surface energy increases as follows; Pt (111) < Pt (110) < Pt (100). The (111) orientation is the energetically favourable growth plane of platinum crystals. Polycrystalline platinum is prone to oxidation due to the symmetric polycrystalline structure. Reactant molecules adsorb strongly on the Pt (111) orientation, hence, it has a lower electrocatalytic activity compared to the Pt (110) and Pt (100) orientations. On the other hand, Pt (111) orientation has a lower poisoning rate than the Pt (110) and Pt (100) orientations [72, 73]. The desired crystal orientations can be obtained by adding a shape-directing agent to modify the crystallisation process. Many researchers have reported that platinum is the best electrocatalyst in acidic media. However, contrary results have been observed when using carbon nanodots as supports. It was observed that the Pt/CNDs electrocatalyst in alkaline media yields more current compared to acidic conditions as can be seen from **Figure 2**.

Gwebu et al. [42] went on to study the electro-oxidation of ethanol on the Pt/CNDs electrode in both acidic and alkaline solutions. A similar trend was observed; high current densities were recorded in alkaline conditions. The results are shown in **Figure 3**.

3.2 Platinum-tin (Pt-Sn) electrocatalyst

Researchers have demonstrated that tin (Sn) enhances the electrocatalytic performance of platinum towards ethanol oxidation and works even better than ruthenium. Among the platinum-based binary catalysts, the Pt-Sn/C electrocatalyst is the most promising for EOR; due its low cost, and high current densities at low potentials. Neto et al. [3] synthesised Pt-Sn/C and Pt-Ru/C by the alcohol reduction process using water and ethylene glycol as the solvent and reducing agent respectively. The activity for methanol and ethanol electro-oxidation was studied at room temperature. The Pt-Sn/C electrocatalyst yielded current values higher than those of Pt-Ru/C for both ethanol and methanol oxidation. The superior activity of the Pt-Sn/C nanocatalyst is brought by the "bifunctional" mechanism where Sn provides oxygen-containing species to react with the CO poisoning intermediates on the platinum sites [74–78].

Spinacé et al. [79] studied the effect of the synthesis method and atomic composition of Pt and Sn. They synthesised Pt-Sn/C nanocatalysts with Pt:Sn ratios of 50:50 and 90:10 by the alcohol reduction method, using ethylene glycol as a reducing agent, and by borohydride reduction. The Pt-Sn/C nanocatalysts prepared by the

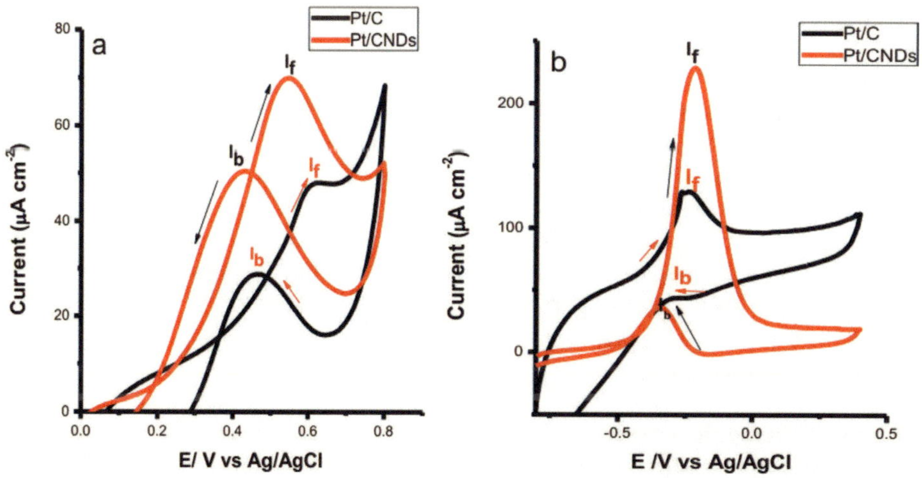

Figure 2.
Comparable CV for Pt/C and Pt/CNDs in (a) 0.5 CH_3OH in 0.5 M H_2SO_4 and (b) 0.5 M CH_3OH in NaOH. Reproduced with permission from Gwebu et al. [42]. Copyright 2017, ESG.

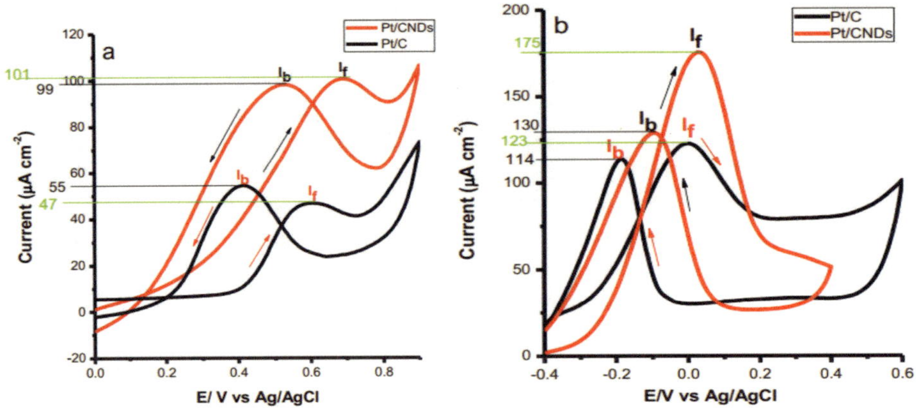

Figure 3.
CV graphs for Pt/C and Pt/CNDs in (a) 3 M CH_3CH_2OH in 0.1 M H_2SO_4 and (b) 2 M CH_3CH_2OH in 0.1 M NaOH. Reproduced with permission from Gwebu et al. [42]. Copyright 2017, ESG.

alcohol reduction method had smaller sizes compared to the Pt-Sn nanocatalysts prepared by borohydride method. The difference in particle sizes was attributed to the stabilising influence of ethylene glycol that minimises nanoparticle growth. The best activity for the electro-oxidation of ethanol was observed for the electrocatalyst containing 50:50 Pt:Sn prepared by the alcohol reduction method. For the borohydride method, higher current densities were observed for the electrocatalyst containing Pt:Sn in the ratio 90:10. Lamy and co-workers [80] demonstrated that Pt-Sn/C nanocatalysts perform better than Pt-Ru/C nanocatalysts for ethanol oxidation. For nanocatalysts synthesized by co-impregnation hydrogen reduction and Bönneman methods, they found that the optimal tin composition was between 10 and 20%.

Zhou and co-workers [81, 82] prepared Pt-Ru/C and Pt-Sn/C nanocatalysts by a polyol method and tested for ethanol oxidation. They observed that the activities of the Pt-Ru/C nanocatalysts were inferior to those of Pt-Sn/C nanocatalysts. They also observed that Pt-Sn/C nanocatalysts with Pt:Sn atomic ratios of 60:40 and 50:50 are more electroactive than nanocatalysts with 75:25 and 80:20 atomic ratios. Some scientists have reported that methanol electro-oxidation is low or insignificant on the Pt-Sn/C electrocatalysts. In actual fact, such observations were later found to be due to other intervening factors than the effect of Sn on methanol oxidation, normally involving "ensemble" effects [83]. Colmati et al. [74] reported that the adsorption-dehydrogenation of methanol turns out to be more difficult due to alloying of Sn with Pt and methanol electro-oxidation occurs only at reasonable alloying ratios.

The performance of Pt-Sn/C nanocatalysts greatly depends on their preparation procedure and composition. Carbon nanodots have been used as support materials for Pt-Sn nanoparticles. It has been found that carbon nanodots supported nanocatalysts show greater electroactivity and slow poisoning rates compared to carbon black supported nanocatalysts as shown in **Figure 4** [84]. The electrochemical activity is usually assessed by cyclic voltammetry and the poisoning rates are normally studied by chronoamperometry.

3.3 Pt-TiO_2 composites

Chemical blending a carbon support material such as a CNTs, CNDs, etc., with TiO_2 helps to improve the anti-poisoning and anti-corrosion properties of the electrode. The synergistic interaction between the carbon support, TiO_2 and the Pt catalyst enhances the electroactivity of the catalyst as it changes the Pt-d electronic and geometric properties [65]. This synergistic effect causes the contraction of Pt-Pt

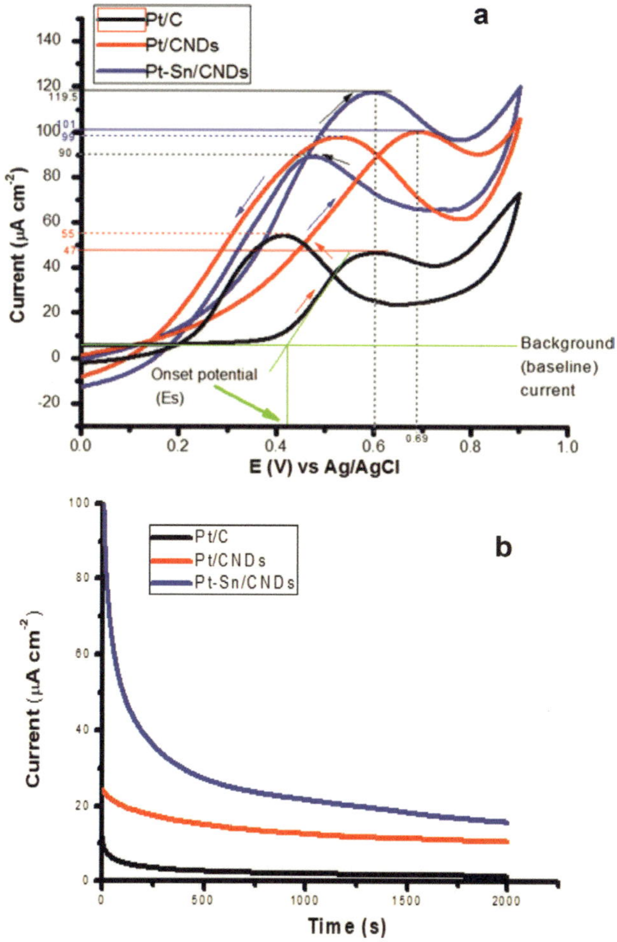

Figure 4.
Cyclic voltammograms showing low onset potentials and high current densities yielded by the Pt-Sn/CNDs nanocatalyst (a), and chronoamperometry curves showing improved poisoning of the Pt-Sn/CNDs nanocatalyst (b). Reproduced with permission from Gwebu et al. [84]. Copyright 2018, Wiley.

bond length leading to a condition favourable for the alcohol electro-oxidation reactions [44]. Titanium dioxide is a semiconductor, hence the quantity of TiO_2 and the crystallography (anatase/rutile) governs the degree of the interaction between the electrocatalyst components [85]. Previous X-ray photoelectron spectroscopic studies reveal that when Pt nanoparticles are supported on CNDs-TiO_2 the electronic structure of Pt is altered by titanium from the composite support. The binding energy for pure Pt_0 is around 71.12 eV, however, upon addition of TiO_2 the binding energy of $4f_{7/2}$ Pt_0 shifts to 71.53 eV. The positive shift indicates a strong metal-support interaction (SMSI) between the support material and the platinum catalyst [86] (**Figure 5**).

The enhanced electrochemical performance provided by metal oxides is brought by their hydrophilic nature due to the availability of H_2O molecules within the oxide network. The water molecules act as a continuous reversible membrane resulting in enhanced hydroxide transfer [43]. Bedolla-Valdez et al. [45] prepared a Pt/CNT/TiO_2 composite for methanol electro-oxidation using the sonochemical method. They concluded that the TiO_2 surface area can offer sites to adsorb water to form hydroxyl groups, which then react with CO adsorbed on the Pt surface to form CO_2. They recommended that the functionalisation of CNT should be optimized. Gwebu et al. [86] prepared a novel Pt/CNDs-TiO_2 nanocatalyst for methanol and ethanol electro-oxidation

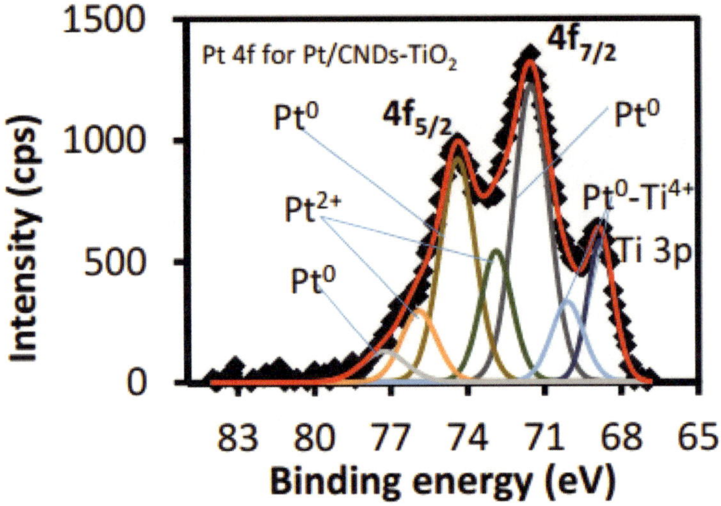

Figure 5.
Pt 4f XPS spectra for Pt/CNDs-TiO$_2$. Reproduced with permission from Gwebu et al. [86]. Copyright 2018, Elsevier.

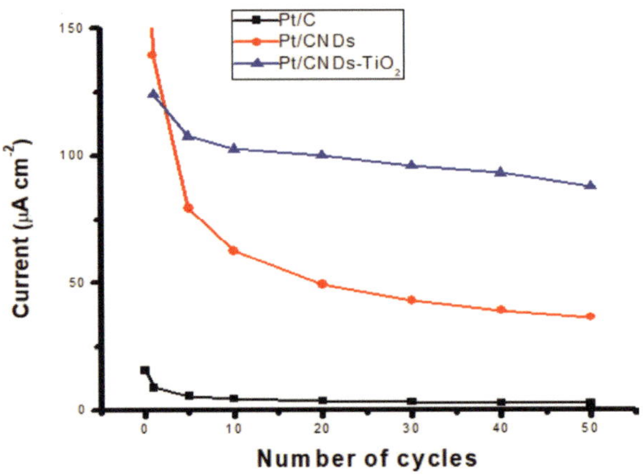

Figure 6.
Stability curves for the Pt/CNDs-TiO$_2$ nanocatalyst. Reproduced with permission from Gwebu et al. [86]. Copyright 2018, Elsevier.

in acidic media. Observed results proved that blending carbon nanodots with TiO_2 not only improves the electroactivity of the nanocatalyst but also enhances the tolerance to poisoning and resistance to corrosion. **Figure 6** shows the durability curves for the Pt/CNDs-TiO_2 nanocatalyst against Pt/CNDs and Pt/C nanocatalysts.

After exposure to 50 cycles, the Pt/CNDs-TiO_2 nanocatalyst retained the highest current indicating great tolerance to poisoning and resistance to acidic conditions of the fuel cell. The improved resistance to poisoning and corrosion was credited to the presence of ceramic TiO_2 nanoparticles which promote the oxidation of poisoning species at the same time minimising the detachment of Pt nanoparticles from the CNDs-TiO_2 composite support.

4. Conclusion

In this chapter, the use of carbon nanodots as primary support materials for direct fuel cell nanocatalysts is discussed. Other carbon nanomaterials such as

CNTs and CNFs were also reviewed. After their discovery in 2004, carbon nanodots have not been extensively used in fuel cells. This chapter demonstrated that carbon nanodots are potential support materials for direct alcohol fuel cells operating in both acidic and alkaline conditions. It is demonstrated that platinum nanoparticles supported on carbon nanodots are better anode nanocatalysts compared to platinum nanoparticles supported on carbon black. Alloying platinum with a cheaper metal such a tin modifies the face centred cubic structure of platinum resulting in a geometry which favours alcohol electrooxidation. The binary Pt-Sn catalyst yielded higher current density at low potentials compared to the mono Pt/CNDs nanocatalyst. It is revealed that incorporating TiO_2 into the Pt/CNDs nanocatalyst improves the nanocatalyst's electroactivity and resistance to corrosion by acidic electrolytes. Chronoamperometry results proved that Pt, Pt-Sn and Pt-TiO_2 nanoparticles supported on carbon nanodots are more resistant to poisoning compared to Pt nanoparticles supported on carbon black.

From the recent studies conducted, it was observed that carbon nanodots are easily synthesized by the bottom-up (physical) methods. However, in most cases CNDs prepared by these methods are amorphous, this compromises their resistance to corrosion under the harsh fuel cell conditions. Further work should be done to develop top-down methods for synthesizing carbon nanodots. Owing to great catalytic activity exhibited by Pt binary catalysts, ternary and quaternary catalysts should be developed to improve performance and reduce platinum loading.

Acknowledgements

This work is based on the research supported wholly/in part by the National Research Foundation of South Africa (Grant Numbers: 112812 and TTK-15071-0125-019). The authors also thank the University of Johannesburg, Faculty of Science, University Research Council for the financial support.

Conflict of interest

No conflicts of interest are declared by the authors.

Author details

Sandile Surprise Gwebu, Philiswa N. Nomngongo and Nobanathi W. Maxakato*
Department of Applied Chemistry, University of Johannesburg, Doornfontein, South Africa

*Address all correspondence to: nmaxakato@uj.ac.za

References

[1] Rong H, Zhang S, Muhammad S, Zhang J. Noble metal-based nanocomposites for fuel cells. In: Kyzas G, editor. Novel Nanomaterials. London, UK: IntechOpen; 2018. pp. 291-310. DOI: 10.5772/intechopen.71949

[2] Kreuer KD. Fuel cells, introduction. In: Kreuer KD, editor. Fuel Cells. New York, NY: Springer; 2013. pp. 1-7. DOI: 10.1007/978-1-4614-5785-5_1

[3] Neto AO, Spinacé EV, Dias RR, Brandalise M, Linardi M. Electro-oxidation of methanol and ethanol using PtRu/C, PtSn/C and PtSnRu/C electrocatalysts prepared by an alcohol-reduction process. Journal of Power Sources. 2007;**116**:87-91. DOI: 10.1016/j.jpowsour.2006.12.088

[4] Lamy C, Belgsir EM, Leger JM. Electrocatalytic oxidation of aliphatic alcohols: Application to the direct alcohol fuel cell (DAFC). Journal of Applied Electrochemistry. 2001;**31**(7):799-809. DOI: 10.1023/A:1017587310150

[5] Soloveichik GL. Liquid fuel cells. Beilstein Journal of Nanotechnology. 2014;**5**:1399-1418. DOI: 10.3762/bjnano.5.153

[6] Zhang L, Wang H, Li X, Xia F, Liu Y, Xu X, et al. One-step synthesis of palladium-gold-silver ternary nanoparticles supported on reduced graphene oxide for the electrooxidation of methanol and ethanol. Electrochimica Acta. 2015;**172**:42-51. DOI: 10.1016/j.electacta.2014.11.152

[7] Park YC, Tokiwa H, Kakinuma K, Watanabe M, Uchida M. Effects of carbon supports on Pt distribution, ionomer coverage and cathode performance for polymer electrolyte fuel cells. Journal of Power Sources. 2016;**315**:179-191. DOI: 10.1016/j.jpowsour.2016.02.091

[8] Dicks AL. The role of carbon in fuel cells. Journal of Power Sources. 2006;**156**(2):128-141. DOI: 10.1016/j.jpowsour.2006.02.054

[9] Trogadas P, Fuller TF, Strasser P. Carbon as catalyst and support for electrochemical energy conversion. Carbon. 2014;**75**:5-42. DOI: 10.1016/j.carbon.2014.04.005

[10] Antolini E, Gonzalez ER. Polymer supports for low-temperature fuel cell catalysts. Applied Catalysis A: General. 2009;**365**(1):1-19. DOI: 10.1016/j.apcata.2009.05.045

[11] Harris PJF. New perspectives on the structure of graphitic carbons. Critical Reviews in Solid State and Materials Sciences. 2005;**30**:235-253. DOI: 10.1080/10408430500406265

[12] Yaldagard M. Carbonaceous nanostructured support materials for low temperature fuel cell electrocatalysts: A review. World Journal of Nano Science and Engineering. 2013;**3**:121-153. DOI: 10.4236/wjnse.2013.34017

[13] Goel J, Basu S. Effect of support materials on the performance of direct ethanol fuel cell anode catalyst. International Journal of Hydrogen Energy. 2014;**39**(28):15956-15966. DOI: 10.1016/j.ijhydene.2014.01.203

[14] Liu J, Liu CT, Zhao L, Zhang JJ, Zhang LM, Wang ZB. Effect of different structures of carbon supports for cathode catalyst on performance of direct methanol fuel cell. International Journal of Hydrogen Energy. 2015;**41**(3):1859-1870. DOI: 10.1016/j.ijhydene.2015.11.103

[15] Jiang ZZ, Wang ZB, Chu YY, Gu DM, Yin GP. Carbon riveted microcapsule Pt/MWCNTs-TiO_2 catalyst prepared by in situ carbonized glucose

with ultrahigh stability for proton exchange membrane fuel cell. Energy & Environmental Science. 2011;**4**(7): 2558-2566. DOI: 10.1039/C1EE01091C

[16] Wang Y, Sheng ZM, Yang H, Jiang SP, Li CM. Electrocatalysis of carbon black- or activated carbon nanotubes-supported Pd-Ag towards methanol oxidation in alkaline media. International Journal of Hydrogen Energy. 2010;**35**(19):10087-10093. DOI: 10.1016/j.ijhydene.2010.07.172

[17] Yang G, Zhou Y, Pan H, Bin ZC, Fu S, Wai CM, et al. Ultrasonic-assisted synthesis of Pd-Pt/carbon nanotubes nanocomposites for enhanced electro-oxidation of ethanol and methanol in alkaline medium. Ultrasonics Sonochemistry. 2016;**28**:192-198. DOI: 10.1016/j.ultsonch.2015.07.021

[18] Wang H, Kakade BA, Tamaki T, Yamaguchi T. Synthesis of 3D graphite oxide-exfoliated carbon nanotube carbon composite and its application as catalyst support for fuel cells. Journal of Power Sources. 2014;**260**:338-348. DOI: 10.1016/j.jpowsour.2014.03.014

[19] Wen Y, Ye JS, Zhang WD, Sheu FS, Xu GQ. Electrocatalytic oxidation of methanol on a platinum modified carbon nanotube electrode. Microchimica Acta. 2008;**162**:235-243. DOI: 10.1007/s00604-007-0882-0

[20] Acosta DR, Martınez L. Growth of carbon nanotubes and nanofibres in porous anodic alumina film. Carbon. 2002;**40**:1011-1016. DOI: 10.1016/S0008-6223(01)00230-5

[21] Prasanna D, Selvaraj V. Development of non-covalent ternary polymer–CNT composites as a novel supporting material for electrooxidation of glycerol. RSC Advances. 2015;**5**(120):98822-98833. DOI: 10.1039/C5RA17172E

[22] Soin N, Roy SS, Sharma S, Thundat T, Mclaughlin JA. Electrochemical and oxygen reduction properties of pristine and nitrogen-doped few layered graphene nanoflakes (FLGs). Journal of Solid State Electrochemistry. 2013;**17**(8):2139-2149. DOI: 10.1007/s10008-013-2073-8

[23] Brouzgou A, Podias A, Tsiakaras P. PEMFCs and AEMFCs directly fed with ethanol: A current status comparative review. Journal of Applied Electrochemistry. 2013;**43**(2):119-136. DOI: 10.1007/s10800-012-0513-2

[24] Murphy H, Papakonstantinou P, Okpalugo TIT. Raman study of multiwalled carbon nanotubes functionalized with oxygen groups. Journal of Vacuum Science & Technology B. 2006;**24**(2):715-720. DOI: 10.1116/1.2180257

[25] Knupp SL, Li W, Paschos O, Murray TM, Snyder J, Haldar P. The effect of experimental parameters on the synthesis of carbon nanotube/nanofiber supported platinum by polyol processing techniques. Carbon. 2008;**6**:1276-1284. DOI: 10.1016/j.carbon.2008.05.007

[26] Guha A, Lu W, Zawodzinski TA, Schiraldi DA. Surface-modified carbons as platinum catalyst support for PEM fuel cells. Carbon. 2007;**45**(7):1506-1517. DOI: 10.1016/j.carbon.2007.03.023

[27] Day TM, Unwin PR, Wilson NR, Macpherson JV. Electrochemical templating of metal nanoparticles and nanowires on single-walled carbon nanotube networks. American Chemical Society. 2005;**127**(20):10639-10647. DOI: 10.1021/ja051320r

[28] Kunitomo H, Ishitobi H, Nakagawa N. Optimized CeO_2 content of the carbon nanofiber support of PtRu catalyst for direct methanol fuel cells. Journal of Power Sources. 2015;**297**:400-407. DOI: 10.1016/j.jpowsour.2015.07.002

[29] Ghouri ZK, Barakat NAM, Obaid M, Lee JH, Kim HY. Co/CeO_2-decorated carbon nanofibers as effective non-precious electro-catalyst for fuel cells application in alkaline medium. Ceramics International. 2015;**41**(2):2271-2278. DOI: 10.1016/j.ceramint.2014.10.031

[30] Ghouri ZK, Barakat NAM, Kim HY, Park M, Khalil KA, El-Newehy MH, et al. Nano-engineered ZnO/CeO_2 dots@CNFs for fuel cell application. Arabian Journal of Chemistry. 2016;**9**(2):219-228. DOI: 10.1016/j.arabjc.2015.05.024

[31] Sharma S, Pollet BG. Support materials for PEMFC and DMFC electrocatalysts: A review. Journal of Power Sources. 2012;**208**:96-119. DOI: 10.1016/j.jpowsour.2012.02.011

[32] Zheng J, Zhang X, Li P, Zhu J, Zhou X, Yuan W. Effect of carbon nanofiber microstructure on oxygen reduction activity of supported palladium electrocatalyst. Electrochemistry Communications. 2007;**9**:895-900. DOI: 10.1016/j.elecom.2006.12.006

[33] Baker S. Luminescent carbon nanodots: Emergent nanolights luminescent carbon nanodots: Emergent nanolights angewandte. Angewandte Chemie, International Edition. 2010;**49**:6726-6744. DOI: 10.1002/anie.200906623

[34] Li H, Kang Z, Liu Y, Lee S-T. Carbon nanodots: Synthesis, properties and applications. Journal of Materials Chemistry. 2012;**22**(46):24230. DOI: 10.1039/C2JM34690G

[35] Roy P, Chen PC, Periasamy AP, Chen YN, Chang HT. Photoluminescent carbon nanodots: Synthesis, physicochemical properties and analytical applications. Materials Today. 2015;**18**(8):447-458. DOI: 10.1016/j.mattod.2015.04.005

[36] Zhu S, Song Y, Zhao X, Shao J, Zhang J, Yang B. The photoluminescence mechanism in carbon dots (graphene quantum dots, carbon nanodots, and polymer dots): Current state and future perspective. Nano Research. 2015;**8**(2):355-381. DOI: 10.1007/s12274-014-0644-3

[37] Wei W, Chen W. "Naked" Pd nanoparticles supported on carbon nanodots as efficient anode catalysts for methanol oxidation in alkaline fuel cells. Journal of Power Sources. 2012;**204**:85-88. DOI: 10.1016/j.jpowsour.2012.01.032

[38] Qin X, Lu W, Asiri AM, Al-Youbi AO, Sun X. Microwave-assisted rapid green synthesis of photoluminescent carbon nanodots from flour and their applications for sensitive and selective detection of mercury(II) ions. Sensors and Actuators B: Chemical. 2013;**184**:156-162. DOI: 10.1016/j.snb.2013.04.079

[39] Sonthanasamy RSA, Ahmad WYW, Fazry S, Hassan NI, Lazim AM. Transformation of crystalline starch nanoparticles into highly luminescent carbon nanodots: Toxicity studies and their applications. Carbohydrate Polymers. 2016;**137**:488-496. DOI: 10.1016/j.carbpol.2015.11.021

[40] Yan F, Kong D, Fu Y, Ye Q, Wang Y, Chen L. Construction of carbon nanodots/tungsten trioxide and their visible-light sensitive photocatalytic activity. Journal of Colloid and Interface Science. 2016;**466**:268-274. DOI: 10.1016/j.jcis.2015.12.043

[41] Shi L, Li Y, Li X, Zhao B, Wen X, Zhang G, et al. Controllable synthesis of green and blue fluorescent carbon nanodots for pH and Cu^{2+} sensing in living cells. Biosensors & Bioelectronics. 2016;77:598-602. DOI: 10.1016/j.bios.2015.10.031

[42] Gwebu SS, Nomngongo PN, Mashazi PN, Nyokong T, Maxakato

NW. Platinum nanoparticles supported on carbon nanodots as anode catalysts for direct alcohol fuel cells. International Journal of Electrochemical Science. 2017;**12**:6365-6378. DOI: 10.20964/2017.07.09

[43] Meenakshi S, Nishanth KG, Sridhar P, Pitchumani S. Spillover effect induced Pt-TiO_2/C as ethanol tolerant oxygen reduction reaction catalyst for direct ethanol fuel cells. Electrochimica Acta. 2014;**135**:52-59. DOI: 10.1016/j.electacta.2014.04.142

[44] Ruiz-Camacho B, Martínez-Álvarez O, Rodríguez-Santoyo HH, Granados-Alejo V. Pt/C and Pt/TiO_2-C electrocatalysts prepared by chemical vapor deposition with high tolerance to alcohols in oxygen reduction reaction. Journal of Electroanalytical Chemistry. 2014;**725**:19-24. DOI: 10.1016/j.jelechem.2014.04.019

[45] Bedolla-Valdez ZI, Verde-Gomez Y, Valenzuela-Muniz AM, Gochi-Ponce Y, Oropeza-Guzman MT, Berhault G, et al. Sonochemical synthesis and characterization of Pt/CNT, Pt/TiO_2, and Pt/CNT/TiO_2 electrocatalysts for methanol electro-oxidation. Electrochimica Acta. 2015;**186**:76-84. DOI: 10.1016/j.electacta.2015.10.084

[46] Odetola C, Trevani L, Easton EB. Enhanced activity and stability of Pt/TiO_2/carbon fuel cell electrocatalyst prepared using a glucose modifier. Journal of Power Sources. 2015;**294**:254-263. DOI: 10.1016/j.jpowsour.2015.06.066

[47] Yang F, Ma L, Gan M, Zhang J, Yan J, Huang H, et al. Polyaniline-functionalized TiO_2–C supported Pt catalyst for methanol electro-oxidation. Synthetic Metals. 2015;**205**:23-31. DOI: 10.1016/j.synthmet.2015.03.017

[48] Zhang J, Tu JP, Du GH, Dong ZM, Su QM, Xie D, et al. Pt supported self-assembled nest-like-porous WO_3 hierarchical microspheres as electrocatalyst for methanol oxidation. Electrochimica Acta. 2013;**88**:107-111. DOI: 10.1016/j.electacta.2012.10.060

[49] Georgieva J, Sotiropoulos S, Valova E, Armyanov S, Karanasios N. Methanol oxidation and photo-oxidation at Pt/WO_3 electrocatalysts on graphite substrates. Journal of Electroanalytical Chemistry. 2014;**727**:135-140. DOI: 10.1016/j.jelechem.2014.06.011

[50] Hernández-Pichardo ML, González-Huerta RG, Del Angel P, Tufiño-Velazquez M, Lartundo L. The role of the WO_3 nanostructures in the oxygen reduction reaction and PEM fuel cell performance on WO_3-Pt/C electrocatalysts. International Journal of Hydrogen Energy. 2015;**40**(48):17371-17379. DOI: 10.1016/j.ijhydene.2015.06.165

[51] Wu M, Han M, Li M, Li Y, Zeng J, Liao S. Preparation and characterizations of platinum electrocatalysts supported on thermally treated CeO_2-C composite support for polymer electrolyte membrane fuel cells. Electrochimica Acta. 2014;**139**:308-314. DOI: 10.1016/j.electacta.2014.07.029

[52] Naeem R, Ahmed R, Ansari MS. TiO_2 and Al_2O_3 promoted Pt/C nanocomposites as low temperature fuel cell catalysts for electro oxidation of methanol in acidic media. IOP Conference Series: Materials Science and Engineering. 2014;**60**:012031. DOI: 10.1088/1757-899X/60/1/012031

[53] Gharibi H, Sadeghi S, Golmohammadi F. Electrooxidation of Ethanol on highly active and stable carbon supported $PtSnO_2$ and its application in passive direct ethanol fuel cell: Effect of tin oxide synthesis method. Electrochimica Acta. 2016;**190**:1100-1112. DOI: 10.1016/j.electacta.2015.12.208

[54] De A, Datta J, Haldar I, Biswas M. Catalytic intervention of MoO_3 toward ethanol oxidation on PtPd nanoparticles decorated MO_3—Polypyrrole composite support. ACS Applied Materials & Interfaces. 2016;**8**(42):28574-28584. DOI: 10.1021/acsami.6b07455

[55] Devi RS, Venckatesh DR, Sivaraj DR. Synthesis of titanium dioxide nanoparticles by sol-gel technique. International Journal of Innovative Research in Science, Engineering and Technology. 2014;**3**(8):15206-15211. DOI: 10.15680/IJIRSET.2014.0308020

[56] Falaras P, Xagas P. Roughness and fractality of nanostructured TiO_2 films prepared via sol-gel technique. Materials Science. 2002;**37**:3855-3860. DOI: 10.1023/A:1019686902277

[57] Shen PS, Tseng CM, Kuo TC, Shih CK, Li MH, Chen P. Microwave-assisted synthesis of titanium dioxide nanocrystalline for efficient dye-sensitized and perovskite solar cells. Solar Energy. 2015;**120**:345-356. DOI: 10.1016/j.solener.2015.07.036

[58] Li W, Zhao Y, Yuan S, Shi L, Wang Z, Fang J, et al. Synthesis and characterization of highly dispersed TiO_2 nanocrystal colloids by microwave-assisted hydrothermal method. Journal of Materials Science. 2012;**47**:7999-8006. DOI: 10.1007/s10853-012-6689-3

[59] Padmamalini N, Ambujam K. Structural and dielectric properties of ZrO_2–TiO_2–V_2O_5 nanocomposite prepared by CO-precipitation calcination method. Materials Science in Semiconductor Processing. 2016;**41**:246-251. DOI: 10.1016/j.mssp.2015.09.009

[60] Zhang X, Sun Y, Cui X, Jiang Z. Carbon-incorporated TiO_2 microspheres: Facile flame assisted hydrolysis of tetrabutyl orthotitanate and photocatalytic hydrogen production. International Journal of Hydrogen Energy. 2012;**37**(2):1356-1365. DOI: 10.1016/j.ijhydene.2011.09.133

[61] Liu J, Hu Y, Gu F, Li C. Large-scale synthesis of hollow titania spheres via flame combustion. Particuology. 2011;**9**(6):632-636. DOI: 10.1016/j.partic.2010.12.004

[62] Piccirillo C, Denis CJ, Pullar RC, Binions R, Parkin IP, Darr JA, et al. Aerosol assisted chemical vapour deposition of hydroxyapatite-embedded titanium dioxide composite thin films. Journal of Photochemistry and Photobiology A Chemistry. 2017;**332**:45-53. DOI: 10.1016/j.jphotochem.2016.08.010

[63] Nguyen TGH, Pham TVA, Phuong TX, Lam TXB, Tran VM, Nguyen TPT. Nano-Pt/C electrocatalysts: Synthesis and activity for alcohol oxidation. Advances in Natural Sciences: Nanoscience and Nanotechnology. 2013;**4**(3):035008/1-035008/8. DOI: 10.1088/2043-6262/4/3/035008

[64] Kuppan B, Selvam P. Platinum-supported mesoporous carbon (Pt/CMK-3) as anodic catalyst for direct methanol fuel cell applications: The effect of preparation and deposition methods. Progress in Natural Science: Materials International. 2012;**22**(6):616-623. DOI: 10.1016/j.pnsc.2012.11.005

[65] Jukk K, Kongi N, Tarre A, Rosental A, Treshchalov AB, Kozlova J, et al. Electrochemical oxygen reduction behaviour of platinum nanoparticles supported on multi-walled carbon nanotube/titanium dioxide composites. Journal of Electroanalytical Chemistry. 2014;**735**:68-76. DOI: 10.1016/j.jelechem.2014.10.008

[66] Chung DY, Lee KJ, Sung YE. Methanol electro-oxidation on the Pt surface: Revisiting the cyclic voltammetry interpretation. Journal of Physical Chemistry C.

2016;**120**(17):9028-9035. DOI: 10.1021/acs.jpcc.5b12303

[67] Wang X, Hu C, Xiong Y, Liu H, Du G, He X. Carbon-nanosphere-supported Pt nanoparticles for methanol and ethanol electro-oxidation in alkaline media. Journal of Power Sources. 2011;**196**(4):1904-1908. DOI: 10.1016/j.jpowsour.2010.09.072

[68] Figueiredo MC, Solla-Gullón J, Vidal-Iglesias FJ, Nisula M, Feliu JM, Kallio T. Carbon-supported shape-controlled Pt nanoparticle electrocatalysts for direct alcohol fuel cells. Electrochemistry Communications. 2015;**55**:47-50. DOI: 10.1016/j.elecom.2015.03.019

[69] Perez J, Paganin VA, Antolini E. Particle size effect for ethanol electro-oxidation on Pt/C catalysts in half-cell and in a single direct ethanol fuel cell. Journal of Electroanalytical Chemistry. 2011;**654**(1-2):108-115. DOI: 10.1016/j.jelechem.2011.01.013

[70] Holton OT, Stevenson JW. The role of platinum in proton exchange membrane fuel cells. Platinum Metals Review. 2013;**57**(4):259-271. DOI: 10.1595/147106713x671222

[71] Barmparis GD, Lodziana Z, Lopez N, Remediakis IN. Nanoparticle shapes by using Wulff constructions and first-principles calculations. Beilstein Journal of Nanotechnology. 2015;**6**(1):361-368. DOI: 10.3762/bjnano.6.35

[72] Antoniassi RM, Otubo L, Vaz JM, Oliveira Neto A, Spinacé EV. Synthesis of Pt nanoparticles with preferential (100) orientation directly on the carbon support for direct ethanol fuel cell. Journal of Catalysis. 2016;**342**:67-74. DOI: 10.1016/j.jcat.2016.07.022

[73] Khudhayer WJ, Shaikh AU, Karabacak T. Platinum nanorod arrays with preferred morphological and crystal properties for oxygen reduction reaction. Advanced Science Letters. 2011;**4**(3):3551-3559. DOI: 10.1166/asl.2011.1867

[74] Colmati F, Antolini E, Gonzalez ER. Pt-Sn/C electrocatalysts for methanol oxidation synthesized by reduction with formic acid. Electrochimica Acta. 2005;**50**(28):5496-5503. DOI: 10.1016/j.electacta.2005.03.030

[75] López-Suárez FE, Bueno-López A, Eguiluz KIB, Salazar-Banda GR. Pt-Sn/C catalysts prepared by sodium borohydride reduction for alcohol oxidation in fuel cells: Effect of the precursor addition order. Journal of Power Sources. 2014;**268**:225-232. DOI: 10.1016/j.jpowsour.2014.06.042

[76] Lim DH, Choi DH, Lee WD, Park DR, Lee HI. The effect of Sn addition on a Pt/C electrocatalyst synthesized by borohydride reduction and hydrothermal treatment for a low-temperature fuel cell. Electrochemical and Solid-State Letters. 2007;**10**(5):B87-B90. DOI: 10.1149/1.2710182

[77] De Souza EA, Giz MJ, Camara GA, Antolini E, Passos RR. Ethanol electro-oxidation on partially alloyed Pt-Sn-Rh/C catalysts. Electrochimica Acta. 2014;**147**:483-489. DOI: 10.1016/j.electacta.2014.09.141

[78] Lopez-Suarez FE, Carvalho-Filho CT, Bueno-Lopez A, Arboleda J, Echavarrea A, Eguiluz KIB, et al. Platinum-tin/carbon catalysts for ethanol oxidation: Influence of Sn content on the electroactivity and structural characteristics. International Journal of Hydrogen Energy. 2015;**40**(37):12674-12686. DOI: 10.1016/j.ijhydene.2015.07.135

[79] Spinacé EV, Vale LAI, Dias RR, Neto AO. PtSn/C electrocatalysts prepared by different methods for direct ethanol fuel cell. Studies in surface science and catalysis. 2006;**162**:617-624. DOI: 10.1016/S0167-2991(06)80960-2

[80] Lamy C, Rousseau S, Belgsir EM, Coutanceau C, Léger JM. Recent progress in the direct ethanol fuel cell: Development of new platinum-tin electrocatalysts. Electrochim Acta. 2004;**49**(22-23 SPEC. ISS):3901-3908. DOI: 10.1016/j.electacta.2004.01.078

[81] Zhou W, Zhou Z, Song S, Li W, Sun G, Tsiakaras P, et al. Pt based anode catalysts for direct ethanol fuel cells. Applied Catalysis B: Environmental. 2003;**46**(2):273-285. DOI: 10.1016/S0926-3373(03)00218-2

[82] Zhou WJ, Song SQ, Li WZ, Sun GQ, Xin Q, Kontou S, et al. Pt-based anode catalysts for direct ethanol fuel cells. Solid State Ionics. 2004;**175**(1-4):797-803. DOI: 10.1016/j.ssi.2004.09.055

[83] Song C, Zhang J. Electrocatalytic oxygen reduction reaction. In: Zhang J, editor. PEM Fuel Cell Electrocatalysts and Catalyst Layers. Springer: London; 2008. pp. 89-134. DOI: 10.1007/978-1-84800-936-3_2

[84] Gwebu SS, Nomngongo PN, Maxakato NW. Pt-Sn nanoparticles supported on carbon nanodots as anode catalysts for alcohol electro-oxidation in acidic conditions. Electroanalysis. 2018;**30**:1125-1132. DOI: 10.1002/elan.201800098

[85] Silva JCM, Buzzo GS, De Souza RFB. Enhanced electrooxidation of ethanol using Pd/C + TiO_2 electrocatalysts in alkaline media. Electrocatalysis. 2015;**6**:86-91. DOI: 10.1007/s12678-014-0224-z

[86] Gwebu SS, Nomngongo PN, Maxakato NW. Pt/CNDs-TiO_2 electrocatalyst for direct alcohol fuel cells. Materials Today Proceedings. 2018;**5**(4):10460-10469. DOI: 10.1016/j.matpr.2017.12.377

Chapter 3

Recent Development of Graphitic Carbon Nitride-Based Photocatalyst for Environmental Pollution Remediation

Mohamad Fakhrul Ridhwan Samsudin, Nurfatien Bacho and Suriati Sufian

Abstract

Globalization today has helped fuel the global socioeconomic growth of the world and reshaping the growth of the industries. While the development had been remarkable, the rapid rise of industrialization had provoked the sustainable chain of diversity which is reflected by rising pollution level, particularly on the water pollution. On account of the cutting edge of water security issue, engineering photocatalytic material remains crucial in finding new ways to combat the challenge of water pollution through photocatalytic pollutants degradation while at the same time acts as the frontlines for energy conversion and environmental protection. To date, graphitic carbon nitride, g-C_3N_4 had emerged as a promising material of interest in photocatalytic application due to its appealing characteristics such as excellent optical properties and high physiochemical and thermal stability. This chapter will comprehensively discuss an insight into the most recent progress in synthesis, properties and the photocatalytic application of g-C_3N_4, particularly in environmental pollution remediation. Special emphasis is also placed on the most recent strategies for enhancing the photocatalytic performance of the g-C_3N_4 photocatalyst. Finally, the future directions and perspectives will be presented.

Keywords: graphitic carbon nitride, photocatalyst, degradation, nanomaterial

1. Introduction

The exponential growth of the industries over the past decade had exerted substantial pressure on sustainability, mainly on the pillars of the environment. While the gap towards an improved well-being brings many intended benefits. The environmental sustainability challenge is also growing at a large scale and complexity, resulting in severe environmental impacts including water pollutions. Within this context, the rise of industrialization alongside with the rapid growth of the global population has been intimately linked with a higher generation of wastewater [1–3]. The statistical glance from United Nations World Water Assessment Programme (2003) has reported that around 2 million tons of industrial sewage and agricultural waste have been discharged into the water bodies every day. Based on the report by the WHO/UNICEF Joint Monitoring Programme, the current statistics reveal that there are approximately 2.1 billion people who are lack access to clean drinking water. Another report by UNESCO (2017) claims that approximately 80% of the wastewater flow is discharged into the ecosystem without any treatment thus contaminating a large portion of the water bodies. In this sense, the excessive release of anthropogenic pollutants originated from industrial use such as the phenolic compound, heavy metals and dyes had resulted in a deterioration in water quality and pose harmful effect on the living organism, which further emphasizes the need to tackle the water pollution issue [4–6].

Engineering photocatalytic material had emerged as a promising technology to bridge the gap between global energy challenge and environmental remediation. Since the pioneering discovery of photocatalytic water splitting by Fujishima and Honda in 1972, photocatalytic material has attracted interdisciplinary attention due to its diverse potential in various discipline such as solar energy conversion, photocatalytic water splitting for hydrogen production and carbon dioxide reduction, organic pollutants degradation and synthesis of organic compounds [7–9]. For the wastewater treatment field, photocatalytic degradation of pollutants is favorable over the conventional method due to its several advantages. This technique does not require non-renewable energy consumption as it exploits the sustainable solar energy [10–12]. Applicable both for gaseous and aqueous treatment, photocatalysis technology reportedly can degrade a wide range of pollutants and toxic compounds without causing any secondary pollutants. Moreover, the photocatalyst can be easily synthesized through various methods from an abundant readily available precursor. The whole process is not only simple to conduct, low in cost and require a relatively short process time, making the method sustainable for wastewater purification in a large scale application.

At presents, there are a various ongoing effort for the development of the sustainable photocatalytic system, with the focused centered on the development of noble metal-free photocatalyst as TiO_2, g-C_3N_4, $BiVO_4$, ZnO, and carbonaceous materials [12–17]. Among these photocatalysts, graphitic carbon nitride (g-C_3N_4) has elicited significant interest as the next generation of the photocatalyst in engineering photocatalytic field for environmental pollutants degradation due to its excellent physiochemical properties [18–21]. The g-C_3N_4 is novel, metal-free photocatalyst with good light absorption properties owing to its medium band-gap energy of 2.7 eV [19, 22, 23]. The polymeric nature of this conjugated materials allows for facile modification of the photocatalyst to improve its optical properties besides permit multiple excitations from absorption of a single photon, both of which are favorable for efficient pollutants degradation. However, the practical application of g-C_3N_4 is still hindered by some of its individual properties such as low visible light utilization, the high recombination rate of photogenerated electron–hole pairs and slow electron transfer which lead to lower photocatalytic performance [22]. Hence, various strategies have been adopted such as energy band engineering, copolymerization with nitrogen precursor and development of heterostructure system in order to overcome the individual drawbacks of pristine g-C_3N_4 [24–26].

In recognition of the great potential of g-C_3N_4 as a promising visible light driven photocatalyst, this chapter is aiming to provide an overview on the most recent related studies on the development of g-C_3N_4 photocatalyst in the environmental pollution remediation. The history and basic principle of photocatalyst system are well explained in order to promote better understanding on the g-C_3N_4. Afterward, the fundamental properties of g-C_3N_4 and the synthesizing techniques are briefly summarized. Next, the current strategies to enhance the overall performance of g-C_3N_4 photocatalyst are explained. Subsequently, the current perspective and future directions of the g-C_3N_4 photocatalyst are included in this chapter.

2. General overview on g-C_3N_4 photocatalyst

The history of engineering photocatalytic material could be traced back to 1972, where photoelectrochemical splitting of water in the presence of TiO_2 and ultraviolet light by Fujishima and Honda had served as the starting point for photocatalytic reaction [27]. Since then, numerous semiconductor-based photocatalysts have been investigated in an attempt to produce a robust photocatalyst for an efficient novel photocatalytic system. Among them, TiO_2, $BiVO_4$, Fe_2O_3 and ZnO had been identified as a potential promising photocatalytic material [28–30]. Although Titanium oxide (TiO_2) had dominated the photocatalytic arena, this photocatalyst suffers from its negative characteristics which hinder the practical exploitation of this material for large scale application [8, 9]. Hence, other potential materials have been tested and explored in the search for robust photocatalyst for large scale application. Among them, g-C_3N_4 had emerged as one of the promising material and become the new research hotspot for various scientific application owing to its excellent features.

2.1 Introduction to g-C_3N_4 photocatalyst

Graphitic carbon nitride is one of the oldest artificial polymer reported back in 1834. Structurally analogous to graphene, this conjugated polymer is a novel, metal-free with a medium band gap of 2.7 eV [31]. Generally, there are several allotropes of C_3N_4 such as α-C_3N_4, β-C_3N_4, pseudocubic C_3N_4, cubic C_3N_4, g-h triazine and g-C_3N_4. However, g-C_3N_4 is considered the most stable form of C_3N_4 under ambient conditions. **Figure 1** illustrates the basic tectonic units to establish the allotropes of the g-C_3N_4 photocatalyst. It was reported that the tri-s-triazine-based g-C_3N_4 photocatalyst was the most stable phases of C_3N_4 at ambient conditions. This postulation was further verified by Kroke et al. [32] with their first-principles density functional theory (DFT) calculations. Meanwhile, only the pseudocubic and g-h triazine phases have direct band gaps while the other phases have indirect band gap energies [32].

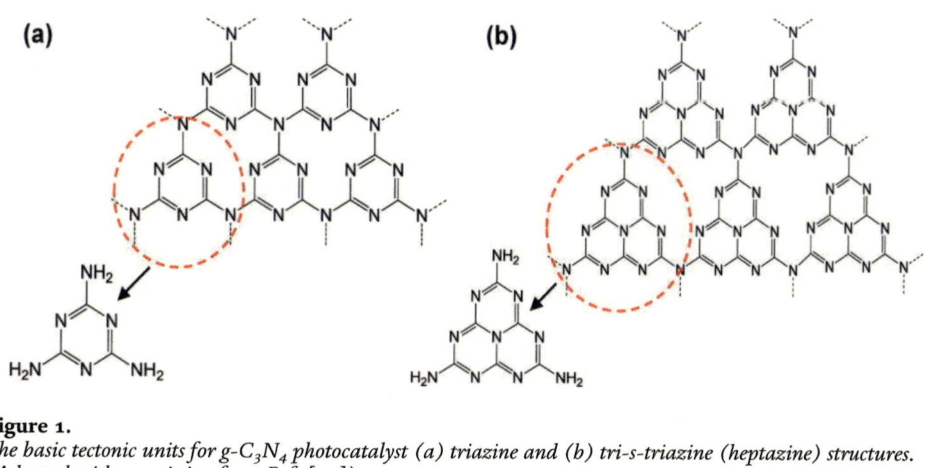

Figure 1.
The basic tectonic units for g-C_3N_4 photocatalyst (a) triazine and (b) tri-s-triazine (heptazine) structures. (Adapted with permission from Ref. [22]).

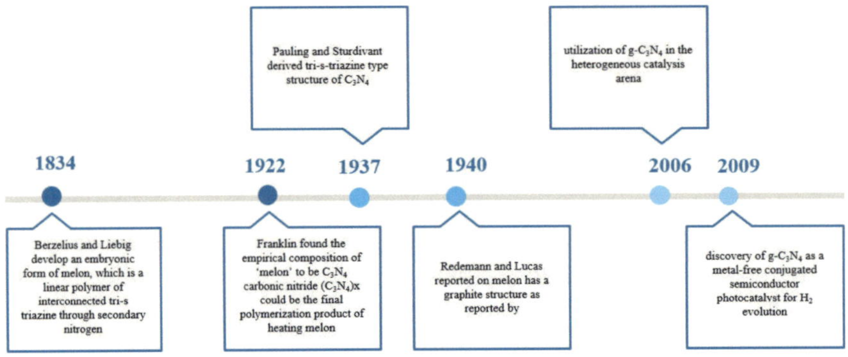

Figure 2.
Historical development of g-C_3N_4 photocatalyst in photocatalysis field.

Carbon nitride, C_3N_4 is not new at all, and it is considered one of the oldest reported artificial polymers in the scientific literature [26]. **Figure 2** shows the historical development of g-C_3N_4 in photocatalysis arena. This g-C_3N_4 was first reported by Berzelius and Liebig in the year 1834 and named as melon [33]. Although it was first discovered back in 1834, the material is not exploited until 2009 when Wang et al. first reported the utilization of this material in photocatalysis field [26]. Since then, a lot of researchers started to unravel the promising potential of the g-C_3N_4 photocatalyst in a wide range of different photocatalytic applications. Nevertheless, owing to the limited experimental data, there is a prevailing discussion about the actual existence of a graphitic material with idealized composition C_3N_4 and possible structure models for g-C_3N_4 photocatalyst. Gratifyingly, due to the similar structure with graphite, triazine (C_3N_4) had been put forward as the elementary block of g-C_3N_4. Moreover, tri-s-triazine rings also shown promises to be energetically favored with respect to the triazine-based modification as the tri-s-triazine rings are cross-linked by trigonal nitrogen atoms [34]. Ideally, the condensed g-C_3N_4 consists of only carbon and nitrogen atoms with a C/N molar ratio of 0.75. Nonetheless, there is no perfectly condensed g-C_3N_4 was reported and they are as-grown polymer materials that are not single crystals. Hence, the g-C_3N_4 photocatalyst can be seen as a family of layered graphitic carbon nitride compounds with a C/N ratio close to 0.75.

2.2 Synthesis method for the development of g-C_3N_4 photocatalyst

Since the potential of carbon nitride in the photocatalysis arena had been known, various synthesis method and technique had been instigated such as chemical vapor deposition, sonochemical, solvothermal, and thermal annealing of nitrogen-rich precursor [24, 32, 33]. Following the variation in synthesis technique, the various compound can be used as the precursor such as melamine, cyanamide, dicyanamide, urea, and thiourea. However, thermal condensation of nitrogen-rich precursor has emerged as the most attractive methods due to its simplicity and use of cheap, earth-abundant precursors.

In the first reported work by Wang et al. [35], cyanamide was used as the starting precursor of g-C_3N_4. It was found that the cyanamide molecules were condensed to dicyanamide and melamine at temperatures of ca. 203 and 234°C. Next, the ammonia was removed via condensation process, resulting in the formation of melamine-based products at the temperature around 335°C. When the temperature was heated up to ca. 390°C, the tri-s-triazine units was formed via rearrangements of melamine. Finally, the polymeric g-C_3N_4 photocatalyst was formed at ca. 520°C through the further condensation of the unit. Nevertheless, the overheating of the sample over 700°C resulted in the disappearance of g-C_3N_4 to "residue-free" through the production of nitrogen and cyano fragments. With respect to the experimental approaches, the reaction mechanism of the combined polyaddition and polycondensation process was further verified by the ab initio calculations using a plane wave basis set with a 550 eV energy cutoff [34]. Based on the calculations, **Figure 3** illustrates the cohesive energy of the molecules increased following the polyaddition pathway, confirming that melamine was produced upon heating the cyanamide [22].

Meanwhile, the most commonly used precursor in thermal condensation method includes those with nitrogen-rich and oxygen-free compounds, which contain the required C-N structure. The compounds with pre-bonded C-N core structure such as melamine, cyanamide, dicyanamide, urea and thiourea are among the commonly used precursors for the synthesis of g-C3N4. In addition to that, triazine and heptazine derivatives also had been tested as a precursor. For example, Mo et al. [34] prepared the g-C_3N_4 photocatalyst via manipulating the calcination temperature on the morphological structure of melamine. It was found that the g-C_3N_4 photocatalyst can only be formed when the calcination temperature above 500°C, evidently from the XRD analysis. Furthermore, it was found that the absorption band edge was red shifted along with the change of color from light yellow to dark orange, indicating the enhanced visible light absorption was obtained for samples with increasing calcination temperatures.

On the other hand, Dong et al. [36] prepared the g-C_3N_4 using urea via a facile template-free. Their group studies the effects of pyrolysis time on the microstructure and activity of g-C_3N_4 photocatalyst. They suggested that the surface areas of the photocatalyst can be significantly increased by just prolonging the pyrolysis time to 240 minutes at under 550°C. They claimed that the surface area of g-C_3N_4 photocatalyst prepared via this method is higher than g-C_3N_4 photocatalyst prepared via templating method. Similarly, Chen et al. [37] prepared g-C_3N_4 photocatalyst by pyrolyzing urea in a muffle furnace at 550°C for 2 hours with a heating rate of 5°C/min. Meanwhile, Yang et al. [38] prepared ultrathin g-C_3N_4 photocatalyst nanosheets via thermal exfoliation of bulk urea-derived g-C_3N_4 under an argon atmosphere. In addition, **Figure 4** summarizes the synthesis process of g-C_3N_4 photocatalyst by thermal polymerization of different precursors.

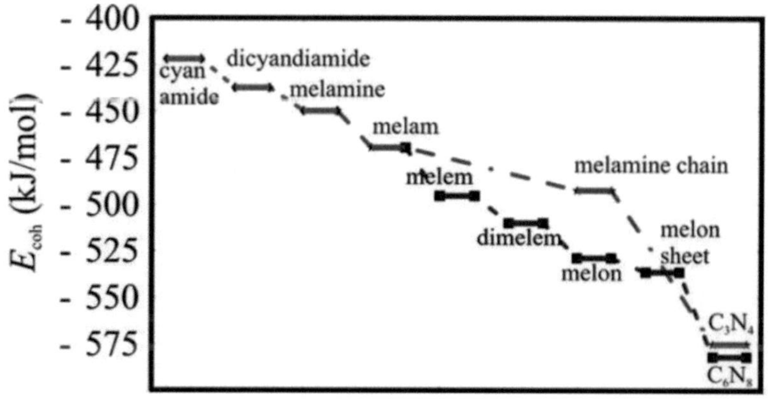

Figure 3.
Calculated energy diagram for the development of C_3N_4 using cyanamide as the precursor. Cyanamide was condensed to melamine. Further condensation proceeded by a triazine route (dash-dot line) or tri-s-triazine (dashed line). (Adapted with permission from Ref. [33]).

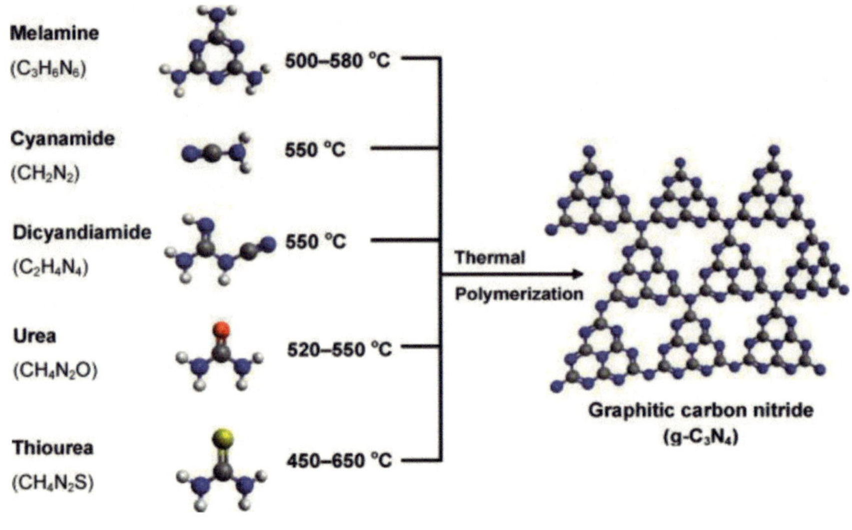

Figure 4.
Schematic illustration of the synthesis process of g-C_3N_4 photocatalyst via thermal polymerization of different precursors. The black, blue, white, red and yellow balls denote C, N, H, O and S atoms, respectively. (Adapted with permission from Ref. [22]).

2.3 Fundamental properties of g-C_3N_4 photocatalyst

Graphitic carbon nitride possesses an excellent physical, chemical and mechanical properties, giving a reason on why researchers are eager into the science of nanotechnology of g-C_3N_4. This compound possesses excellent visible light absorption than most of the metal oxide photocatalyst owing to its mild band gap energy. Given its π-conjugated properties, g-C_3N_4 can act as an electron sink, leading to suppression of recombination of photo excited charge carriers. Moreover, the polymeric nature of this material allows for multiple excitations from absorption of a single photon, leading to an efficient generation of the reactive species responsible for pollutant degradation. In addition, g-C_3N_4 possesses high resistance to thermal and chemical environment, as well as too acidic and basic media, making it as a stable material.

Furthermore, Wang et al. [39] performed a density functional theory (DFT) calculations in order to gain insight into the electronic structure of g-C_3N_4 photocatalyst (as shown in **Figure 5**). They reported that the valence band and conduction band are mainly composed of the nitrogen p_z orbitals and carbon p_z orbitals, respectively. The light illumination excited the electrons and holes for the oxidation and reduction process to occur independently in the nitrogen atoms and carbon atoms. Congruously, the g-C_3N_4 photocatalyst has a specific microstructure, with surface termination as defects and nitrogen atoms for electron localization or anchoring inorganic/organic functional motifs as the active sites [32].

However, the practical application of g-C_3N_4 is still hindered by some of its undesirable properties which lead to lower photocatalytic performance. The individual structure of g-C_3N_4 has low specific surface area and quantum efficiency, which limit the sorption capacity of the photocatalyst [26]. This photocatalyst also suffers from high recombination rate of the photogenerated electron–hole pairs.

2.4 Photocatalytic principles and mechanism over g-C_3N_4 photocatalyst

Photocatalyst can be described as a combination of catalysis and photochemistry in which absorption of photon energy from light via catalyst is the

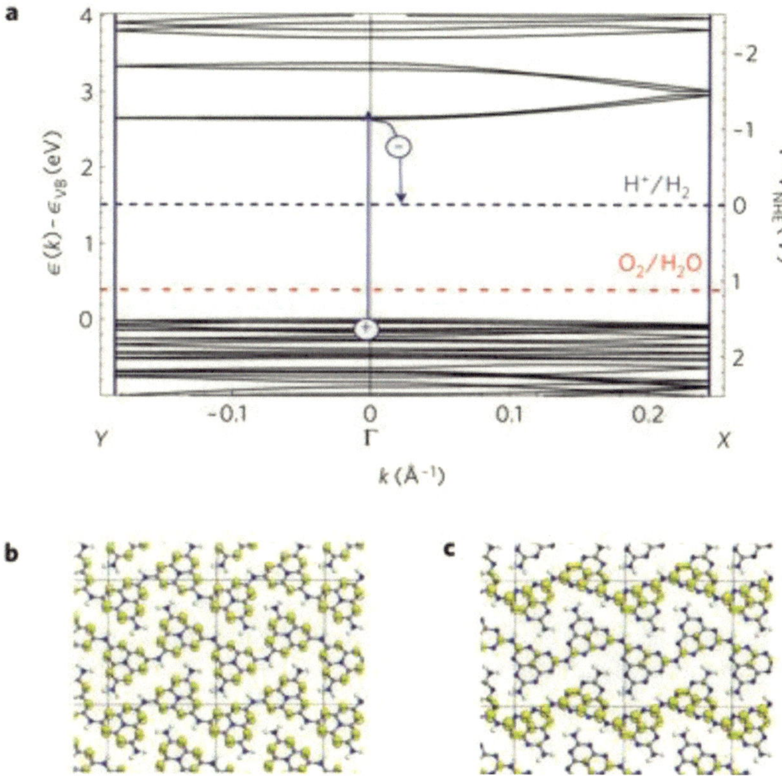

Figure 5.
Electronic structure of the polymeric g-C_3N_4 photocatalyst. (Reproduced with permission from Ref. [39]).

key towards photoreaction process. In this system, photon energy is required to activate the photocatalyst in which make this photocatalyst system to be one of the frontier renewable energy technology in which it can utilize the solar energy. The g-C_3N_4 photocatalyst has a small band gap energy around 2.6–2.7 eV which falls within the visible-light region [25]. It is estimated that the valence band (VB) and the conduction band (CB) of the g-C_3N_4 photocatalyst are 1.56 and 1.09 eV, respectively [40].

Initially, the g-C_3N_4 photocatalyst will absorb photon energy with an energy equivalent or greater than its band gap energy, causing an electron (e^-) in the valence band (VB) to be excited and migrate to the conduction band (CB).

$$g - C_3N_4 + hV \rightarrow e^-{}_{CB}(g - C_3N_4) + h^+{}_{VB}(g - C_3N_4) \quad (1)$$

Then, the electron will be excited and leave photogenerated holes (h^+) in the valence band. The photogenerated holes and photoexcited electrons will migrate to the surface of the photocatalyst and trapped there. The photogenerated holes then react with adsorbed water to produce strong oxidizing $^{\cdot}OH$ radicals whereas the photoexcited electrons react with adsorbed oxygen to generate $^{\cdot}O_2^-$ radicals [41].

$$e^-{}_{CB}(g - C_3N_4) + O_2 \rightarrow \bullet O_2^- \quad (2)$$

$$h^+{}_{VB}(g - C_3N_4) + OH^- \rightarrow \bullet OH \quad (3)$$

The formation of these radicals will further react with recalcitrant pollutant and subsequently degraded the recalcitrant pollutant.

$$\bullet O_2^- + \text{recalcitrant pollutants} \rightarrow \text{degradation of pollutant} \quad (4)$$

$$\bullet OH + \text{recalcitrant pollutants} \rightarrow \text{degradation of pollutant} \quad (5)$$

2.5 Strategies to improve g-C_3N_4 photocatalytic performance

In order to overcome the individual drawback of pure g-C_3N_4, many attempts have been made to improve the photocatalytic capability including surface modification of the photocatalyst structure. Generally, the surface modification aims to improve the photocatalyst specific surface area, charge separation and optical. There are currently five modification techniques which have been investigated including the introduction of heteroatoms (i.e. metals and non-metals) within g-C_3N_4 framework, noble metal deposition, hybridizing g-C_3N_4 with carbon nanomaterials and coupling g-C_3N_4 with a photocatalyst. The principle, advantages and disadvantages of each technique are summarized in **Table 1**.

On the other hand, the development of the heterostructure photocatalyst via the introduction of additional compounds into the g-C_3N_4 network is one of the most promising strategies to enhance the overall catalytic performance of g-C_3N_4 photocatalyst [22]. In general, the g-C_3N_4-based heterostructure can be developed by coupling the g-C_3N_4 with other types of photocatalyst as co-catalyst. The formation of the heterostructure with suitable band position would suppress the recombination rate of the photogenerated electron–hole pairs, which lead to higher

Modification method	Principle	Advantages	Disadvantages
Metal doping	Doping of various metallic species such as the alkali metals, rare earth metals and noble metals into g-C_3N_4	Bandgap narrowing, surface area improvement, charge separation and fine-tuning the band structure	Can often cause secondary pollution due to leaching of the metal ions
Non-metal doping	Doping g-C_3N_4 with non-metals	No secondary pollution, improve visible light absorption and charge separation	Non-metal species does not take part in charge transportation hence recombination centers are formed
Noble metal deposition	Deposition of noble metal nanoparticles such as Cu, Pt, Au and Pd on g-C_3N_4	Metal content positively influence the photocatalytic activity until the optimum loading is reached	Beyond the optimum metal loading, the excess metal ions act as recombination centers for the electron/hole pairs
Hybridizing g-C_3N_4 with carbon nanomaterials (CNM)	Carbon nanomaterial such as carbon nanotubes (CNTs), carbon nanospheres (CNS), graphene oxide (GO) and reduced graphene oxide (RGO)	High thermal, electronic conductivity, remarkable adsorption properties for organic and inorganic compounds	Excess CNM (i.e. RGO) facilitate adsorption of large amounts of the dye molecules onto the catalyst surface thereby reducing light penetration to the photocatalyst
Coupling g-C_3N_4 with semiconductor	Coupling two or more semiconductors to form a semiconductor heterojunction	Improved stability, visible light utilization, charge separation and transfer and more efficient formation of the oxidizing species	Difficult to find a proper semiconductor photocatalyst with suitable band edge position

Table 1.
Surface modification of g-C_3N_4 technique to improve its photocatalytic performance.

photocatalytic performance efficiency. For example, Wang et al. [19] reported the enhanced photocatalytic performance of g-C_3N_4/{010} facets $BiVO_4$ photocatalyst fabricated via ultrasonic dispersion method. The aforementioned heterostructure photocatalyst was capable of removing 88.3% of RhB pollutant within 30 minutes under visible light irradiation. Meanwhile, Huo et al. [20] successfully formed the $BiVO_4$/Polydopamine/g-C_3N_4 heterostructure photocatalyst via facile ultrasonic dispersion and self-assembly at the room temperature. They observed a remarkable photocatalytic degradation of glyphosate under visible light irradiation in comparison to the unmodified photocatalyst.

Furthermore, the photocatalytic activity of pure photocatalyst can be enhanced by the addition of carbon materials as an electron bridge mediator. The addition of carbon nanomaterial provides a structure with a larger specific surface area over which the active component can be well-dispersed, thus increasing the active sites. During the photocatalytic degradation of organic pollutants, carbon materials can be used as an adsorbent to improve the adsorption capacity of semiconductors. Besides, carbon materials can be doped as a photosensitizer for band gap narrowing, which is favorable for expanding the visible light absorption region of semiconductors. The incorporation of electron bridge mediator within the network of the heterostructure system will facilitate the migration of the electron transfer within the photocatalyst, leading to the enhanced charge separation efficiency and photocatalytic activity. In this sense, GO/RGO and CNTs are among the carbon nanomaterial that has been explored to acts as an electron bridge mediator.

3. Recent progress on the development of g-C_3N_4 photocatalyst for environmental remediation

Photocatalyst	Synthesized Method	Type of pollutants	Pollutants concentration	Source of light	Degradation efficiency	Ref.
ZnO/g-C_3N_4	Impregnation	Phenol	5 mg/L	Simulated sunlight irradiation	99.5% of phenol removal achieved after 60 min of irradiation	[42]
$BiVO_4$/ PDA/g-C_3N_4	Ultrasonic dispersion and self-assembly	Glyphosate	0.1 mM	125 W mercury lamp	100% degradation after 150 minutes	[20]
g-C_3N_4/ $BiVO_4$	NR	Methylene blue	10 mg/L	Visible light	g-C_3N_4/$BiVO_4$ photocatalyst was about 8 and 7 times higher than that of pure $BiVO_4$ and g-C_3N_4	[43]
g-C_3N_4/TiO_2/ CNT	Facile hydrothermal method	MB	6 mg/L	150 W halogen lamp	Ternary hybrid exhibits 5 times higher photocatalytic activity compared to bare g-C_3N_4	[44]

Photocatalyst	Synthesized Method	Type of pollutants	Pollutants concentration	Source of light	Degradation efficiency	Ref.
$g\text{-}C_3N_4$/AC	NR	Phenol	50 mg/L	350WXe lamp	100% phenol removal after 160 min	[45]
$g\text{-}C_3N_4$	Polycondensation	Pharmaceutical	20 mg/L 10 mg/L	35 W Xenon lamp	Photocatalyst showed the capacity to degrade the pollutants in the sequence: tetracycline > ciprofloxacin > salicylic acid > ibuprofen	[46]
$g\text{-}C_3N_4$/ CNTs/Au	Ultra-sonication	Aqueous Rhodamine (RhB)	250 mL	Visible light	100 vol% Au dispersion with a 60: 40 mass ratio of g-C3N4 to CNTs exhibited optimal photocatalytic activity	[47]

4. Conclusions and future directions

The growing concern over the scarcity of clean water sources due to a fast development of industrialization has force a rapid breakthrough dedicated to the development of advanced photocatalyst system. Over the past few years, the studies on $g\text{-}C_3N_4$ based photocatalyst have witnessed auspicious potential promises by this photocatalyst in environmental remediation applications. To date, the profound photocatalytic performance of $g\text{-}C_3N_4$ based photocatalyst is mainly governed by its intrinsic features such as metal-free photocatalyst and good light absorption properties owing to its medium band-gap energy of 2.7 eV. In this chapter, the synthesis, properties and photocatalytic application of $g\text{-}C_3N_4$ are summarized. Then, the most recent strategies for enhancing the photocatalytic performance of the $g\text{-}C_3N_4$ photocatalyst are highlighted.

Although profound performance had been reported in most of the recent studies, the promising potential of $g\text{-}C_3N_4$ based photocatalyst has yet to be exploited fully. The main challenges which are yet to be mitigated are (i) green synthesizing method which can produce high surface area and good photostability photocatalyst, (ii) the control of surface kinetics on $g\text{-}C_3N_4$ photocatalyst which can promote the photocharge separation and migration, (iii) the use of real industrial wastewater in analyzing the performance of $g\text{-}C_3N_4$ based photocatalyst, (iv) improving the reactor design to achieve the optimum photocatalytic performance with the lowest cost and (v) the utilization of real sunlight as a light source during the analysis process.

Acknowledgements

The authors would like to express their appreciation to the Chemical Engineering Department, Universiti Teknologi PETRONAS, Centre of Innovative

Nanostructures & Nanodevices (COINN), Universiti Teknologi PETRONAS and The Murata Science Foundation Grant (015ME0-033) for the financial and laboratory support. The authors would like to thank the Centre Analytical Lab, Universiti Teknologi PETRONAS for the sample characterization facilities.

Conflict of interest

The authors declare no conflict of interest.

Author details

Mohamad Fakhrul Ridhwan Samsudin, Nurfatien Bacho and Suriati Sufian*
Chemical Engineering Department, Universiti Teknologi Petronas,
Bandar Seri Iskandar, Perak, Malaysia

*Address all correspondence to: suriati@utp.edu.my

References

[1] Regmi C, Kshetri YK, Kim T, Prasad R, Kumar S, Wohn S. Fabrication of Ni-doped $BiVO_4$ semiconductors with enhanced visible-light photocatalytic performances for wastewater treatment. Applied Surface Science. 2017;**413**:253-265. DOI: 10.1016/j.apsusc.2017.04.056

[2] Wang D, Li Y, Puma GL, Lianos P, Wang C, Wang P. Photoelectrochemical cell for simultaneous electricity generation and heavy metals recovery from wastewater. Journal of Hazardous Materials. 2017;**323**:681-689. DOI: 10.1016/j.jhazmat.2016.10.037

[3] Brillas E, Martínez-Huitle CA. Decontamination of wastewaters containing synthetic organic dyes by electrochemical methods. An updated review. Applied Catalysis B: Environmental. 2015;**166–167**:603-643. DOI: 10.1016/j.apcatb.2014.11.016

[4] Boczkaj G, Fernandes A. Wastewater treatment by means of advanced oxidation processes at basic pH conditions: A review. Chemical Engineering Journal. 2017;**320**:608-633. DOI: 10.1016/j.cej.2017.03.084

[5] Eryuruk K, Tezcan U, Bak U. Electrochemical treatment of wastewaters from poultry slaughtering and processing by using iron electrodes. Journal of Cleaner Production. 2018;**172**: 1089-1095. DOI: 10.1016/j.jclepro.2017.10.254

[6] Arantes MK, Alves HJ, Sequinel R, Da Silva EA. Treatment of brewery wastewater and its use for biological production of methane and hydrogen. International Journal of Hydrogen Energy. 2017;**42**:26243-26256. DOI: 10.1016/j.ijhydene.2017.08.206

[7] Samsudin MFR, Sufian S, Mohamed NM, Bashiri R, Wolfe F, Ramli RM. Enhancement of hydrogen production over screen-printed $TiO_2/BiVO_4$ thin film in the Photoelectrochemical cells. Materials Letters. 2018;**211**:13-16. DOI: 10.1016/j.matlet.2017.09.013

[8] Samsudin MFR, Sufian S, Bashiri R, Mohamed NM, Ramli RM. Synergistic effects of pH and calcination temperature on enhancing photodegradation performance of m-$BiVO_4$. Journal of Taiwan Institute of Chemical Engineering. 2017;**81**:305-315. DOI: 10.1016/j.jtice.2017.09.045

[9] Samsudin MFR, Mahmood A, Sufian S. Enhanced photocatalytic degradation of wastewater over RGO-$TiO_2/BiVO_4$ photocatalyst under solar light irradiation. Journal of Molecular Liquids. 2018;**268**:26-36. DOI: 10.1016/j.molliq.2018.05.012

[10] Bora LV, Mewada RK. Visible/solar light active photocatalysts for organic effluent treatment: Fundamentals, mechanisms and parametric review. Renewable and Sustainable Energy Reviews. 2017;**76**:1393-1421. DOI: 10.1016/j.rser.2017.01.130

[11] Petala A, Bontemps R, Spartatouille A, Frontistis Z, Antonopoulou M, Konstantinou I, et al. Solar light-induced degradation of ethyl paraben with $CuO_x/BiVO_4$: Statistical evaluation of operating factors and transformation by-products. Catalysis Today. 2017;**280**: 122-131. DOI: 10.1016/j.cattod.2016.04.045

[12] Wang F, Wang Y, Feng Y, Zeng Y, Xie Z, Zhang Q, et al. Novel ternary photocatalyst of single atom-dispersed silver and carbon quantum dots co-loaded with ultrathin g-C_3N_4 for broad spectrum photocatalytic degradation of naproxen. Applied Catalysis B: Environmental. 2018;**221**:510-520. DOI: 10.1016/j.apcatb.2017.09.055

[13] Guo Q, Zhang Q, Wang H, Zhao Z. ZnO_2-promoted ZnO as an efficient photocatalyst for the photoreduction of carbon dioxide in the presence of water. Catalysis Communications. 2018;**103**: 24-28. DOI: 10.1016/j.catcom.2017. 09.010

[14] Liu Y, Kong J, Yuan J, Zhao W, Zhu X, Sun C, et al. Enhanced photocatalytic activity over flower-like sphere $Ag/Ag_2CO_3/BiVO_4$ plasmonic heterojunction photocatalyst for tetracycline degradation. Chemical Engineering Journal. 2018;**331**:242-254. DOI: 10.1016/j.cej.2017.08.114

[15] Ou M, Wan S, Zhong Q, Zhang S, Song Y, Guo L, et al. Z-scheme photocatalyst of $g\text{-}C_3N_4$@Ag/$BiVO_4$ (040) with enhanced visible-light-induced photocatalytic oxidation performance. Applied Catalysis B: Environmental. 2018;**221**:97-107. DOI: 10.1016/j.apcatb.2017.09.005

[16] Kumar S, Jain S, Yadav Lamba B, Kumar P. Epigrammatic status and perspective of sequestration of carbon dioxide: Role of TiO_2 as photocatalyst. Solar Energy. 2018;**159**:423-433. DOI: 10.1016/j.solener.2017.11.007

[17] Ong CB, Ng LY, Mohammad AW. A review of ZnO nanoparticles as solar photocatalysts: Synthesis, mechanisms and applications. Renewable Sustainaible Energy Review. 2018;**81**: 536-551. DOI: 10.1016/j.rser.2017. 08.020

[18] Zhang Y, Wang M, Yang QY, Chai T, Li S, Zhu T. Novel binary of $g\text{-}C_3N_4$ coupling and Eu^{3+} doping co-modifying bidirectional dendritic $BiVO_4$ heterojunctions with enhanced visible-light photocatalytic performance. Separation and Purification Technology. 2018;**202**: 335-344. DOI: 10.1016/j.seppur.2018. 04.013

[19] Wang Y, Tan G, Liu T, Su Y, Ren H, Zhang XL, et al. Photocatalytic properties of the $g\text{-}C_3N_4$/{010} facets $BiVO_4$ interface Z-scheme photocatalysts induced by $BiVO_4$ surface heterojunction. Applied Catalysis B: Environmental. 2018;**234**: 37-49. DOI: 10.1016/j.apcatb.2018. 04.026

[20] Huo R, Yang XL, Yang JY, Yang SY, Xu YH. Self-assembly synthesis of $BiVO_4$ /Polydopamine/$g\text{-}C_3N_4$ with enhanced visible light photocatalytic performance. Materials Research Bulletin. 2018;**98**:225-230. DOI: 10.1016/ j.materresbull.2017.10.016

[21] Xiao P, Jiang D, Ju L, Jing J, Chen M. Construction of RGO/$CdIn_2S_4$/$g\text{-}C_3N_4$ ternary hybrid with enhanced photocatalytic activity for the degradation of tetracycline hydrochloride. Applied Surface Science. 2018;**433**:388-397. DOI: 10.1007/ s10853-015-9388-z

[22] Ong WJ, Tan LL, Ng YH, Yong ST, Chai SP. Graphitic carbon nitride ($g\text{-}C_3N_4$)-based Photocatalysts for artificial photosynthesis and environmental remediation: Are we a step closer to achieving sustainability? Chemical Reviews. 2016;**116**:7159-7329. DOI: 10.1021/acs.chemrev.6b00075

[23] Sun J, Guo Y, Wang Y, Cao D, Tian S, Xiao K, et al. H_2O_2 assisted photoelectrocatalytic degradation of diclofenac sodium at $g\text{-}C_3N_4$/$BiVO_4$ photoanode under visible light irradiation. Chemical Engineering Journal. 2018;**332**:312-320. DOI: 10.1016/j.cej.2017.09.041

[24] Xu B, Ahmed MB, Zhou JL, Altaee A, Xu G, Wu M. Graphitic carbon nitride based nanocomposites for the photocatalysis of organic contaminants under visible irradiation: Progress, limitations and future directions. The Science of the Total Environment. 2018;

633:546-559. DOI: 10.1016/j.scitotenv.2018.03.206

[25] Jiang L, Yuan X, Pan Y, Liang J, Zeng G, Wu Z, et al. Doping of graphitic carbon nitride for photocatalysis: A reveiw. Applied Catalysis B: Environmental. 2017;**217**:388-406. DOI: 10.1016/j.apcatb.2017.06.003

[26] Masih D, Ma Y, Rohani S. Graphitic C_3N_4 based noble-metal-free photocatalyst systems: A review. Applied Catalysis B: Environmental. 2017;**206**:556-588. DOI: 10.1016/j.apcatb.2017.01.061

[27] Fujishima A, Honda K. Electrochemical photolysis of water at a semiconductor electrode. Nature. 1972; **238**:37-38

[28] Chong MN, Jin B, Chow CWK, Saint C. Recent developments in photocatalytic water treatment technology: A review. Water Research. 2010;**44**:2997-3027. DOI: 10.1016/j.watres.2010.02.039

[29] Abe R. Recent progress on photocatalytic and photoelectrochemical water splitting under visible light irradiation. Journal of Photochemistry and Photobioliology C: Photochemistry Reviews. 2010;**11**: 179-209. DOI: 10.1016/j.jphotochemrev.2011.02.003

[30] Kudo A. Recent progress in the development of visible light-driven powdered photocatalysts for water splitting. International Journal of Hydrogen Energy. 2007;**32**:2673-2678. DOI: 10.1016/j.ijhydene.2006.09.010

[31] Zhang J, Li Y, Zhu P, Huang D, Wu S, Cui Q, et al. Graphitic carbon nitride materials synthesized via reactive pyrolysis routes and their properties. Diamond and Related Materials. 2011; **20**:385-388. DOI: 10.1016/j.diamond.2011.01.028

[32] Zheng Y, Lin L, Wang B, Wang X. Graphitic carbon nitride polymers toward sustainable Photoredox catalysis. Angewandte Chemie, International Edition. 2015;**54**:12868-12884. DOI: 10.1002/anie.201501788

[33] Thomas A, Fischer A, Goettmann F, Antonietti M, Müller JO, Schlögl R, et al. Graphitic carbon nitride materials: Variation of structure and morphology and their use as metal-free catalysts. Journal of Materilas Chemistry. 2008;**18**: 4893-4908. DOI: 10.1039/b800274f

[34] Gong Y, Li M, Li H, Wang Y. Graphitic carbon nitride polymers: Promising catalysts or catalyst supports for heterogeneous oxidation and hydrogenation. Green Chemistry. 2015; **17**:715-736. DOI: 10.1039/c4gc01847h

[35] Wang X, Chen X, Thomas A, Fu X, Antonietti M. Metal-containing carbon nitride compounds: A new functional organic-metal hybrid material. Advanced Materials. 2009;**21**:1609-1612. DOI: 10.1002/adma.200802627

[36] Dong F, Wang Z, Sun Y, Ho WK, Zhang H. Engineering the nanoarchitecture and texture of polymeric carbon nitride semiconductor for enhanced visible light photocatalytic activity. Journal of Colloid and Interface Science. 2013;**401**:70-79. DOI: 10.1016/j.jcis.2013.03.034

[37] Chen Q, Hou H, Zhang D, Hu S, Min T, Liu B, et al. Enhanced visible-light driven photocatalytic activity of hybrid $ZnO/g\text{-}C_3N_4$ by high performance ball milling. Journal of Photochemistry and Photobiology, A: Chemistry. 2018;**350**:1-9. DOI: 10.1016/j.jphotochem.2017.09.015

[38] Yang Y, Chen J, Mao Z, An N, Wang D, Fahlman BD. Ultrathin $g\text{-}C_3N_4$ nanosheets with an extended visible-light-responsive range for significant enhancement of photocatalysis. RSC

Advances. 2017;7:2333-2341. DOI: 10.1039/C6RA26172H

[39] Wang X, Maeda K, Thomas A, Takanabe K, Xin G, Carlsson JM, et al. A metal-free polymeric photocatalyst for hydrogen production from water under visible light. Nature Materials. 2009;**8**:76-80. DOI: 10.1038/nmat2317

[40] Liang Q, Jin J, Liu C, Xu S, Yao C, Chen Z, et al. Hydrothermal fabrication of α-$SnWO_4$/g-C_3N_4 heterostructure with enhanced visible-light photocatalytic activity. Journal materials science: Materials in. Electronics. 2017; **28**:11279-11283. DOI: 10.1007/s10854-017-6918-2

[41] Samsudin MFR, Sufian S, Hameed BH. Epigrammatic progress and perspective on the photocatalytic properties of $BiVO_4$-based photocatalyst in photocatalytic water treatment technology: A review. Journal of Molecular Liquids. 2018;**268**:438-459. DOI: 10.1016/j.molliq.2018.07.051

[42] Md Rosli NI, Lam SM, Sin JC, Satoshi I, Mohamed AR. Photocatalytic performance of ZnO/g-C_3N_4 for removal of phenol under simulated sunlight irradiation. Journal of Environmental Engineering. 2018;**144**: 1-12. DOI: 10.1061/(ASCE)EE.1943-7870.0001300

[43] Cheng J, Yan X, Mo Q, Liu B, Wang J, Yang X, et al. Facile synthesis of g-C_3N_4/$BiVO_4$ heterojunctions with enhanced visible light photocatalytic performance. Ceramics International. 2017;**43**:301-307. DOI: 10.1016/j.ceramint.2016.09.156

[44] Chaudhary D, Vankar VD, Khare N. Noble metal-free g-C_3N_4/TiO_2/CNT ternary nanocomposite with enhanced photocatalytic performance under visible-light irradiation via multi-step charge transfer process. Solar Energy. 2017;**158**:132-139. DOI: 10.1016/j.solener.2017.09.012

[45] Chen X, Kuo DH, Lu D. Nanonization of g-C_3N_4 with the assistance of activated carbon for improved visible light photocatalysis. RSC Advances. 2016;**6**:66814-66821. DOI: 10.1039/C6RA10357J

[46] Hernández-Uresti DB, Vázquez A, Sanchez-Martinez D, Obregón S. Performance of the polymeric g-C_3N_4 photocatalyst through the degradation of pharmaceutical pollutants under UV-vis irradiation. Journal of Photochemistry and Photobiology, A: Chemistry. 2016;**324**:47-52. DOI: 10.1016/j.jphotochem.2016.01.031

[47] Pawar RC, Kang S, Ahn SH, Lee CS. Gold nanoparticle modified graphitic carbon nitride/multi-walled carbon nanotube (g-C_3N_4/CNTs/Au) hybrid photocatalysts for effective water splitting and degradation. RSC Advances. 2015;**5**:24281-24292. DOI: 10.1039/c4ra15560b

Chapter 4

Platinum Group Metal Based Nanocatalysts for Environmental Decontamination

Sarre M.K. Nzaba, Bhekie B. Mamba and Alex T. Kuvarega

Abstract

Research and development in chemical engineering is currently focused on design of highly active and selective catalytic systems for process intensification. In recent years, there has been growing interest in the use of catalysts based on nanosized metal particles to improve catalytic processes. Among the many metal catalysts, platinum group metals (PGMs) have received greater attention because of their physical and catalytic properties. They have found applications in a wide range of chemical conversion and environmental decontamination reactions due to their chemical stability and enhanced catalytic reactivity in the nano range. This chapter reviews some of the major innovative applications of PGM nanocatalysts for catalytic environmental decontamination.

Keywords: nanocatalyst, decontamination, platinum group metal, organic transformation, water treatment

1. Introduction

Industrialization and rapid population growth has resulted in energy shortages and environmental contamination which has raised concern of a potential global crisis. For sustainable human society development, technologies for environmental decontamination need urgent attention. Among numerous available technologies, catalysis has gained considerable attention because of the diverse potentials in energy and environmental applications. Generally, catalysts for environmental applications are based on less expensive materials that will not cause secondary environmental pollution [1]. The major advantage of environmental catalysis is the chemical conversion of pollutants into non-hazardous and less toxic products. Pollutants can be degraded and transformed efficiently through homogenous or heterogeneous oxidation and reduction processes under ambient conditions or conditions in which external energy such as light may be required [2]. Therefore, the present book chapter aims to provide analysis in the use of PGMs nanocatalysts in the recent development and appraise their potential applications in environmental decontamination.

1.1 Nanotechnology and the environment

Nanotechnology refers to the research and development of materials at the atomic, molecular or macromolecular scale. Materials at nanoscale find applications in a myriad of areas, such as magnetic and optoelectronic, biomedical, pharmaceutical, cosmetic, energy, electronic, catalytic, and environmental domains. Because of the potential of nanotechnology, there has been a worldwide increase in investment in nanotechnology research and development [3]. The unique properties of materials and their stupendous performance at nanoscale are the main reason for the increased growth in this area. Controlled assembly of nanoparticles has been proposed as one of the most ways to achieve the target technologies [4]. Nanotechnology has immense potential in environmental decontamination through the use of materials such as adsorbents and nanocatalysts. Therefore, it is necessary to develop novel processes for the fabrication of nanomaterials that can be used as the basis for the development of highly efficient new technologies for solving environment challenges.

1.2 Platinum group metals

Platinum group metals (PGMs) including iridium (Ir), osmium (Os), platinum (Pt), palladium (Pd), rhodium (Rh) and ruthenium (Ru) have high resistance to corrosion and oxidation in moist air, unlike most base metals [5]. Their precious nature derives from their rarity in Earth's crust. It has been reported by Grabowska et al., that PGMs such as platinum and palladium allow the extension of light absorption of semiconductors such as TiO_2 into the visible region [6].

Recently, Dozzi et al. has investigated the catalytic effect of PGMs metal for the degradation of formic acid using Pt, Au doped TiO_2 [7]. On the other hand, Kisch et al. modified TiO_2 with chloride complexes of Pt for the photocatalytic degradation of 4-chlorophenol [8]. The presence of PGMs nanocatalysts gave rise to an enhanced charge carrier separation hence improving the photocatalytic performances of the modified photocatalyst systems. According to Yoon et al., the surface of PGMs serves as visible light absorbing sensitizers and centers of charge separation [9]. The possibility of designing and applying nanosized PGM catalysts has been demonstrated in a number of studies [9–12]. This has resulted in interest in considering their application in environmental remediation, water treatment, chemical transformation and microbial disinfection. A priority task of present day science is to solve the problem of environmental contamination of planetary resources which requires development of efficient technologies employing nanocatalysts. PGMs catalysts have already found applications in a number of industrially important R&D niche areas, including environmental cleanup. The effectiveness of the PGM nanocatalysts depends on a number of factors, including their size, morphology of the particles and their packaging into usable devices and systems. In most of these applications, the cost of PGMs cannot be ignored. Platinum has been reported to be a very good catalyst but a trade-off has to be established between efficacy and cost. In the nano-range the amount of the catalyst is usually kept very low while huge improvements in efficiency are realized. However, there is still need for more effort in design of viable systems for environmental decontamination using PGM based nanocatalysts. Potential up-scaling of these devices and systems is currently an area of increasing research interest.

2. Synthesis of PGM nanocatalysts

Designing a new class of highly selective and active catalytic systems with the use of recent developments in chemistry has become one of the main concerns faced in contemporary engineering. As compared to bulk material, nanoparticles have an increased surface area and high dispersity and, thus provide high reactivity and allow fabrication of efficient catalysts with lower noble metal loading. Controlling the growth, size, and monodispersity of metal nanoparticles is a subject of interest in designing of nanosized catalytic systems. Therefore, different methods have been used for the synthesis of nanocatalysts. For instance, in 2005, Wong et al., successfully synthesized a bimetallic catalyst using Pd-on-Au NPs through the Turkevich–Frens (citrate reduction) method and obtained particles with an average diameter of about 20 nm [10]. Paula and co-workers synthesized Pd/C-catalyst using one-pot method for synthesis of secondary amines by hydrogenation of nitrocompounds as single starting materials [11]. Additionally, Coleman et al. synthesized and Pt/TiO_2 catalyst by the photodeposition method in the presence of a sacrificial organic hole scavenger [12]. Barakat et al., synthesized Pt doped TiO_2 catalyst by immobilizing colloidal Pt nanoparticles onto titanium dioxide (rutile) [13]. Kuvarega et al., used a modified sol gel method for the synthesis of nitrogen, PGM co doped TiO_2. In these studies spherical particles of average size between 2 and 5 nm were reported [14].

3. Characterization of PGMs based nanocatalysts

Many different physico-chemical techniques are used to characterize nanocatalysts among them scanning electron microscopy (SEM) used for the morphology of the materials. Transmission electron microscopy (TEM) is also used for particles size and morphology. Chun-Hua et al. used TEM to analyze 3% Pt nanoparticles supported on CNT particles is shown in **Figure 1**. In another study, Pt catalyst containing 3% metal

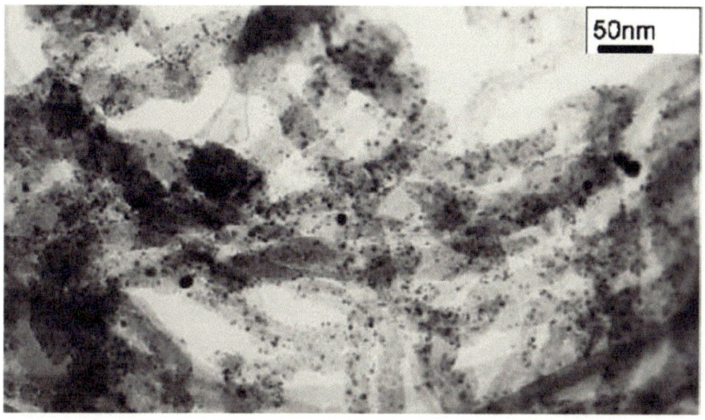

Figure 1.
TEM image of 3% Pt/CNT particles (Copyright, J. Mol. Catal. A: Chem., Ref. [17]).

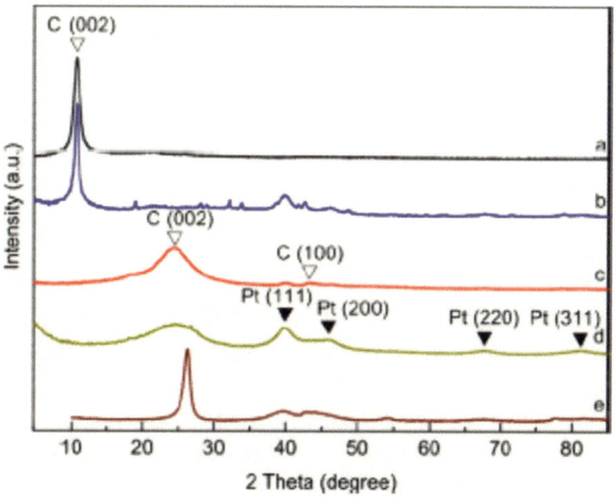

Figure 2.
XRD patterns of (a) GO, (b) Pt/RGO-HH, (c) Pt/RGO-EG, (d) Pt/AC-EG and (e) Pt/MWCNT-EG. (Copyright Carbon 50 (2012) 586-596, Ref [18]).

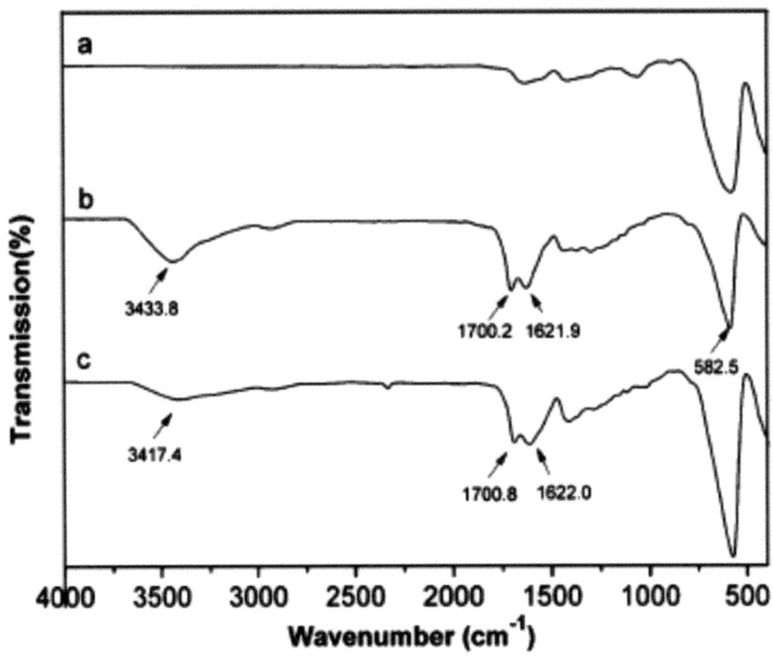

Figure 3.
FT-IR spectra of (a) the Fe_3O_4 microspheres, (b) the Fe_3O_4@C composite, (c) the Fe_3O_4@C@Pt catalyst (reproduced with permission from the Royal Society of Chemistry).

supported on activated carbon (AC) was prepared and the particle size varied from 8 to 10 nm [15]. In addition, techniques such as X-Ray Photoelectron Spectroscopy (XPS) have been used to give information on the oxidation states of the PGM and the nature of bonding between the metals and the supports.

Raman spectroscopy (RS) and X-ray diffraction (XRD) are conducted to identify the crystalline phases and estimate particle sizes of nanocatalysts. For instance, Renfeng et al. used XRD to identify the Pt in Pt/RGO [16]. The Pt peak was conspicuous in the samples containing Pt (**Figure 2**).

Xie and co-workers used Fourier Transform infrared spectroscopy (FTIR) to verify the bond vibrations related to functionalities on the surfaces of their synthesized materials (**Figure 3**) [17].

4. Application of PGMs based nanocatalysts

4.1 Environmental decontamination

PGMs nanocatalysts have become a new class of environmental remediation materials that could provide affordable solutions to some of the environmental challenges. At nanoscale, PGMs particles have high surface areas and surface reactivity. As such, they provide more flexibility for in situ applications. PGMs nanocatalysts have proven to be an excellent choice for the transformation and decontamination of a wide variety of common environmental contaminants, such as organochlorine pesticides and chlorinated organic compounds. The major consumer of PGMs is the automobile industry. PGMs such as Pt, Rh and Pd are used as catalysts in the automobile industry in order to reduce the level of unburnt hydrocarbons, carbon monoxide (CO) and nitrogen oxide present in the exhaust gases. Generally, a typical automobile converter contains 0.04% Pd, 0.005–0.007%

and 0.08% Pt and Rh supported on a base [18]. Iridium has become the new entrant in this application area. A typical example is the introduction of iridium-containing catalytic converters in their direct injection engines by Mitsubishi of Japan [19].

A typical example is the reduction of nitrogen oxides to nitrogen used in car exhaust systems for abatement of emissions from petrol/rich-burn engines. There is a huge number of reports on use of Rhodium as it is the most effective element in the conversion of nitrous oxides to nitrogen [20, 21].

Jianbing et al., in their study, investigated the catalytic ozonation of dimethyl phthalate (DMP) in aqueous solution and DBP precursors in natural water using Ru/AC. These two kinds of organics are both recalcitrant to biodegradation and will cause severe hazards to human health.

Ru/AC was an active nanocatalysts in the catalytic ozonation of dimethyl phthalate and had the ability to complete mineralize the DMP in a semi-batch experiment. On the other hand, the total organic carbon (TOC) removals were stable around 75% for a duration of 42 h and no trace of Ru was observed from the reactor in the continuous experiments of Ru/AC catalyzed ozonation of DMP. Consequently, Ru/AC catalyzed ozonation was found to be more efficient than ozonation alone for TOC removals in the natural water treatment [22].

4.2 Water treatment

Oxyanions, such as BrO_3^- ClO_3^-, NO_3^- and ClO_4^- are toxic and they are pervasive in drinking water. These oxyanions originate from both anthropogenic and natural sources. Furthermore, they are also produced during water treatment processes such as ozonation, desalination, electrochemical treatment and chlorination [23–26]. These ions have mutagenic, endocrine disrupting and carcinogenic effects [27, 28]. Ion exchange and reverse osmosis cannot completely degrade these oxyanions [28, 29]. Thus, it would be preferable to apply destructive treatment technologies based on nanocatalysts for sustainable drinking water treatment processes [30]. Pd-based heterogeneous catalysis has garnered momentous attention as a potential solution for reduction of these oxyanions and for other highly oxidized contaminants such as halogenated and nitro organics [31]. Chen et al., have used PGMs for the reduction of BrO_3^- [32]. Five activated carbon supported on metal namely (with a 5 wt % Pd, Pt, Rh, Ru and 1 wt% for Ir) with a M/C of 0.1 g L catalysts loading were used for the catalytic reduction of bromated (BrO_3^-) **Figure 4**. From **Figure 4**, it is noticeable that Rh/C was significantly more active than the other PGMs/C catalysts used for the reduction of BrO_3^-.

When catalyst loading 0.1 g L^{-1} was used, a reduction of 1 mM BrO_3^- was achieved in approximately 5 minutes. Rh/C had a much higher activity than most supported metal catalyst reported in literature when compared with metal mass- normalized basis. While each metal catalyst dispersion differs, it was vital to compare the activity of metals hydrogenation using the initial turnover frequency values (TOF_o) [33–35]. Additionally, Rh/C had the best performance compared to the other four catalysts at pH 7.2. On the other hand, Ir/C had a slightly lower apparent reactivity with BrO_3^- than that of Pd/C but was the second highest performance. Therefore, the incorporation of the two PGMs catalysts namely Rh and Ir led to the highest activity for the catalytic reduction of $BrO3^-$ under condition that are suitable for the water treatment systems.

4.3 Antimicrobial

Microbial contamination and growth on the surfaces are risks to human health. The chemicals used to tackle microbials such as detergents, alcohols and chlorine are very aggressive and hence not environmentally friendly besides being

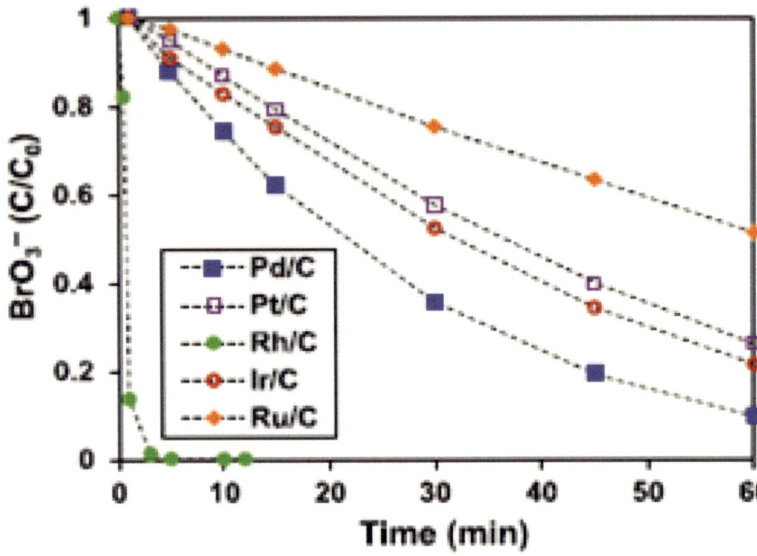

Figure 4.
Kinetics of 1 mM BrO_3^- by 0.1 g L^{-1} M/C catalyst at 1 atm H_2, pH 7.2 and 22°C (with 5 wt% metal for Pd, Rh, Ru, and Pt; 1 wt% metal for Ir). (Copyright Chemical Engineering Journal, 313, 2017, 745-752, Ref. [28]).

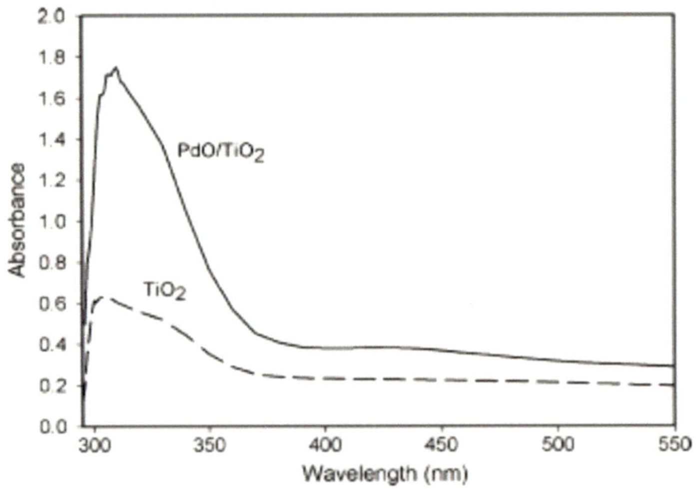

Figure 5.
UV–vis spectra of bare TiO_2 and PdO/TiO_2 samples. (Copyright, Chemistry 184 (2006) [34]).

ineffective for long-term disinfection. Therefore, a myriad of studies are being conducted in order to tackle these challenges. Arcan et al. doped SnO_2 and TiO_2 with Pd for microbial inactivation of *E. coli, S. aureus* and *S. cerevisiae* [36].

The addition of Pd led to an enhancement in the photocatalytic efficiency observed for the degradation of microorganisms when 1% of Pd was used. In addition, the UV–Vis showed an extension of the absorption edge into the visible range without affection the phase of the catalysts (**Figures 5** and **6**).

4.4 Chemical transformation

PGM nanoparticles have proven to be efficient heterogeneous and homogeneous catalyst with advantages such as a high specific surface area due to their small

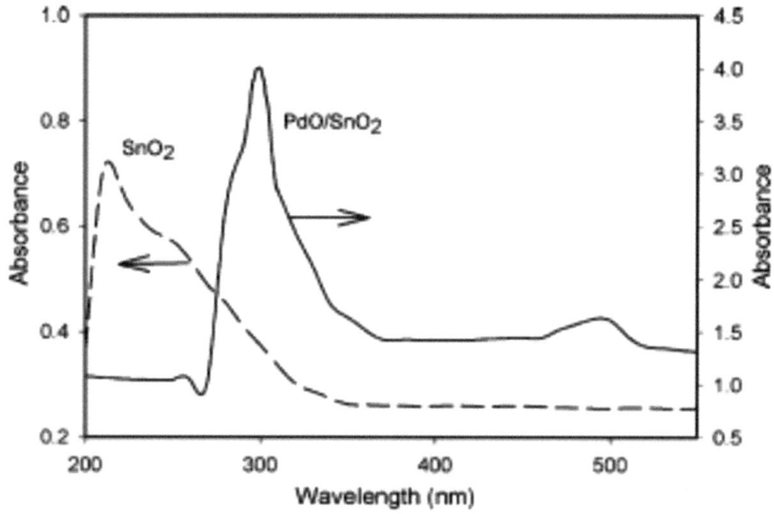

Figure 6.
UV–vis spectra of bare SnO_2 and PdO/SnO_2 samples. (Copyrights Chemistry 184 (2006) Ref. [34]).

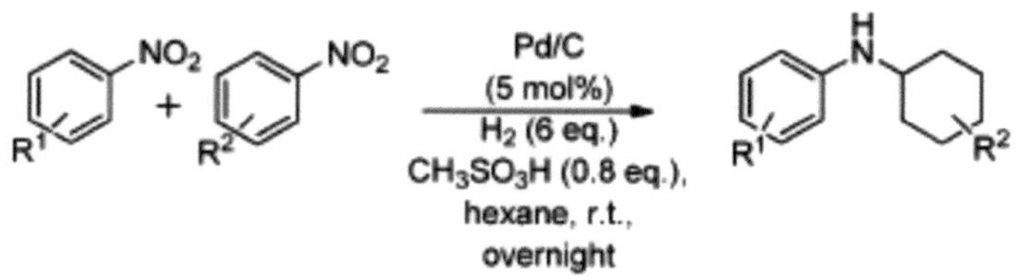

Figure 7.
Pd/C-catalyzed one-pot synthesis of secondary amines by hydrogenation of nitrocompounds. (Copyright Chem. Commun., 2013, 49, 8160, Ref [10]).

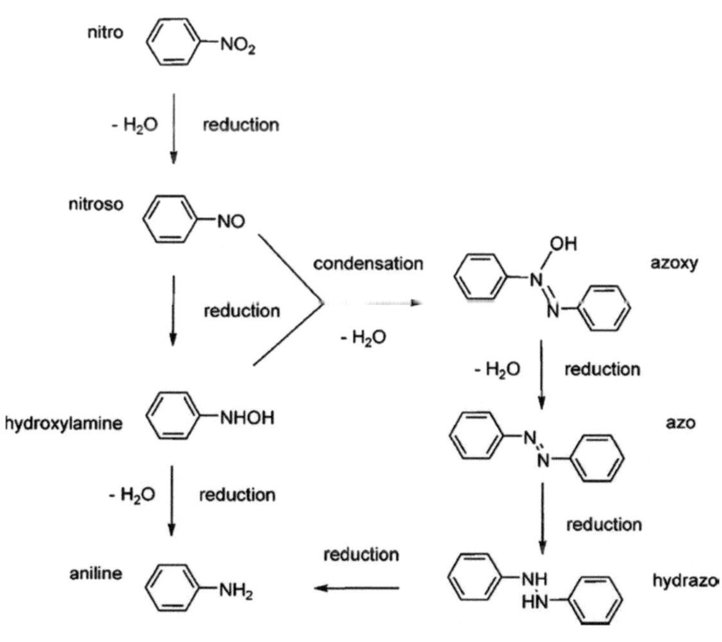

Figure 8.
Proposed mechanism for the hydrogenation of nitroarenes. (Copyright ChemCatChem, 2009, 1, 210–221, Ref. [38]).

Entry	Substrate	Product	t/min	Conv./%	Sel./%
1[a]	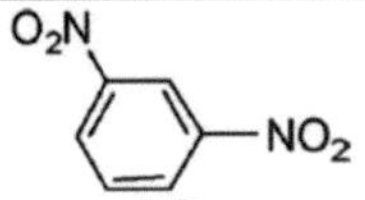	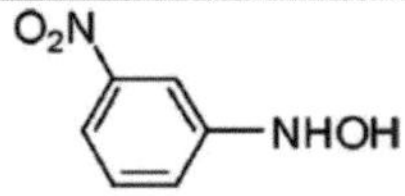	190	100	92.3
2[b]	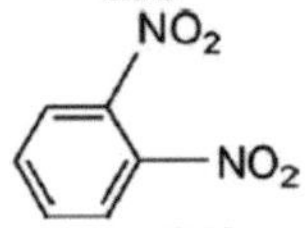	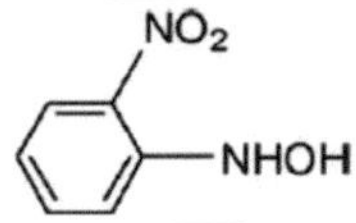	170	100	91.9
3[b]		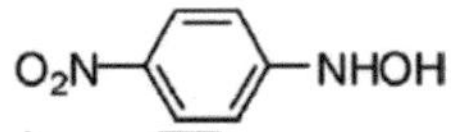	115	100	86.3
4[b]	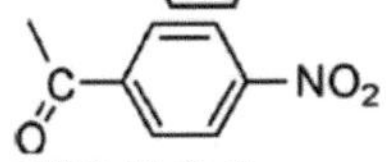	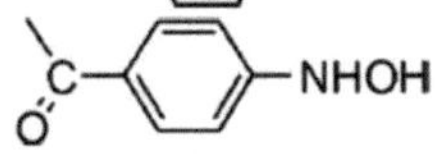	65	100	86.3
5[c]	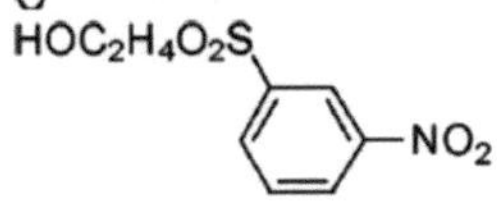	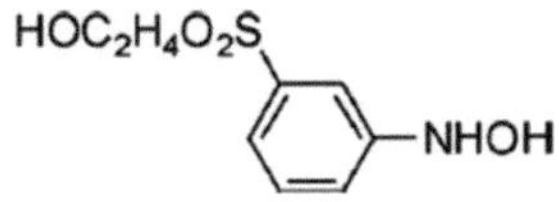	210	100	93.4

Table 1.
Hydrogenations performed using Pt/C as catalysts (Catal. Sci. Technol., 2014, 2445, Ref. [39]).

size resulting in a high number of potential catalytic sites [37]. Owing to these phenomena, there has been in the past two decades an increase in the use of metal nanoparticles in catalysis [38, 39]. Rong et al., reported the synthesis of supported Pt NPs and their use as catalysts in the partial hydrogenation of nitroarenes to arylhydroxylamines [40]. The particles were prepared by reduction of H_2PtCl_6 with $NaBH_4$ in the presence of the carbon support. The hydrogenation of several substituted nitroarenes was performed under soft conditions (10.15°C, 1 bar H_2) to favor the formation of hydroxylamines, showing excellent activity and selectivity in this transformation. Pd supported on C has also been successfully used as catalysts for the conversion of nitrobenzenes to secondary amines (**Figure 7**).

The proposed mechanism for the hydrogenation of nitrobenzene was proposed as shown in **Figure 8**. During this process, there is generation of intermediates such as hydroxylamines, azo and azoxy derivatives.

For example, in the case of m-dinitrobenzene, a catalyst containing 2 wt% Pt/C yielded 92.3% of the corresponding hydroxylamine after 190 min of reaction in THF (**Table 1**).

In a study, Zeming et al. used carbon as Pt colloid support for the hydrogenation of arylhydroxylamines. The Pt colloid supported on carbon was an active and selective catalyst for the partial hydrogenation of nitroaromatics with electron-withdrawing substituents to the corresponding N-arylhydroxylamine, indicating an additive-free green catalytic approach for arylhydroxylamine synthesis. Very encouraging results were obtained with N-arylhydroxylamine bearing electron-withdrawing substituents. Since N-arylhydroxylamine can be further converted to highly valuable compounds through several reactions like Bamberger rearrangement, this result will generally contribute to a simpler and greener synthetic methodology of N-arylhydroxylamine derivatives.

5. Future perspectives

Progress has been reported in the application of nano-PGMs for heterogeneous catalysis reactions. While most of the applications have centered on organic transformations, there is potential for extending the catalytic potential of these metals to other fields such as pollutant degradation and microbial inactivation in water treatment processes. On their own, PGM at the nanoscale tend to aggregate and thus limited application has been realized. However, the use of supports has greatly enhanced the activity of the PGM nanocatalysts in various fields. Mono and bimetallic systems have been reported on. In environmental decontamination processes there is still need to find the most suitable supports and application devices. While encouraging findings have started appearing in literature, more work still needs to be done to in environmental catalysis for water treatment.

6. Conclusion

Several efforts have been devoted to the preparation of PGM nanocatalysts for application in environment decontamination. The high number of literature reports highlights the interest in this family of catalysts for catalytic transformation, both in terms of reactivity and selectivity. There is potential for application of PGMs nanocatalysts for environmental decontamination, water treatment, antimicrobial and chemical transformation. Further studies are necessary to better understand parameters influencing the reactivity as well as enhancing the conversion rates and efficiencies.

Acknowledgements

Appreciation towards funding received from the University of South Africa (UNISA), the National Research Fund (NRF) and support from the Nanotechnology and Water Sustainability and Research Unit (NanoWS) is highly expressed.

Conflict of interest

The authors declare that there is no conflict of interests concerning the publication of this book chapter.

Author details

Sarre M.K. Nzaba, Bhekie B. Mamba and Alex T. Kuvarega*
Nanotechnology and Water Sustainability Research Unit, College of Science, Engineering and Technology, University of South Africa, Florida Campus, Johannesburg, South Africa

*Address all correspondence to: kuvarat@unisa.ac.za

References

[1] Zhang Y et al. TiO_2–graphene nanocomposites for gas-phase photocatalytic degradation of volatile aromatic pollutant: Is TiO_2–graphene truly different from other TiO_2–carbon composite materials? ACS Nano. 2010;**4**(12):7303-7314

[2] Barber J, Tran PD. From natural to artificial photosynthesis. Journal of the Royal Society Interface. 2013;**10**(81):20120984

[3] Dunphy Guzmán KA, Taylor MR, Banfield JF. Environmental risks of nanotechnology: National Nanotechnology Initiative funding, 2000–2004. Environmental Science & Technology. 2006;**40**(5):1401-1407

[4] Biswas P, Wu C-Y. Nanoparticles and the environment. Journal of the Air & Waste Management Association. 2005;**55**(6):708-746

[5] Teh CM, Mohamed AR. Roles of titanium dioxide and ion-doped titanium dioxide on photocatalytic degradation of organic pollutants (phenolic compounds and dyes) in aqueous solutions: A review. Journal of Alloys and Compounds. 2011;**509**(5):1648-1660

[6] Grabowska E, Remita H, Zaleska A. Photocatalytic activity of TiO_2 loaded with metal clusters. Physicochemical Problems of Mineral Processing. 2010;**45**:29-38

[7] Dozzi MV, Saccomanni A, Selli E. Cr (VI) photocatalytic reduction: Effects of simultaneous organics oxidation and of gold nanoparticles photodeposition on TiO_2. Journal of Hazardous Materials. 2012;**211**:188-195

[8] Kisch H. Semiconductor photocatalysis—Mechanistic and synthetic aspects. Angewandte Chemie International Edition. 2013;**52**(3):812-847

[9] Yoon J-W. Dispersion of nanosized noble metals in $ TiO_2 $ matrix and their photoelectrode properties. Journal of the Korean Crystal Growth and Crystal Technology. 2009;**19**(5):251-255

[10] Wong MS et al. Cleaner water using bimetallic nanoparticle catalysts. Journal of Chemical Technology & Biotechnology: International Research in Process, Environmental & Clean Technology. 2009;**84**(2):158-166

[11] Rubio-Marqués P, Leyva-Pérez A, Corma A. A bifunctional palladium/acid solid catalyst performs the direct synthesis of cyclohexylanilines and dicyclohexylamines from nitrobenzenes. Chemical Communications. 2013;**49**(74):8160-8162

[12] Coleman HM, Chiang K, Amal R. Effects of Ag and Pt on photocatalytic degradation of endocrine disrupting chemicals in water. Chemical Engineering Journal. 2005;**113**(1):65-72

[13] Barakat M et al. Pt nanoparticles/TiO_2 for photocatalytic degradation of phenols in wastewater. Environmental Technology. 2014;**35**(2):137-144

[14] Kuvarega AT, Krause RW, Mamba BB. Comparison between base metals and platinum group metals in nitrogen, M codoped TiO_2 (M=Fe, Cu, Pd, Os) for photocatalytic removal of an organic dye in water. Journal of Nanomaterials. 2014;**2014**:199

[15] Li C-H et al. Nitrobenzene hydrogenation with carbon nanotube-supported platinum catalyst under mild conditions. Journal of Molecular Catalysis A: Chemical. 2005;**226**(1):101-105

[16] Nie R et al. Platinum supported on reduced graphene oxide as a catalyst for hydrogenation of nitroarenes. Carbon. 2012;**50**(2):586-596

[17] Xie M et al. Pt nanoparticles supported on carbon coated magnetic microparticles: An efficient recyclable catalyst for hydrogenation of aromatic nitro-compounds. RSC Advances. 2013;**3**(26):10329-10334

[18] Balcerzak M. Analytical methods for the determination of platinum in biological and environmental materials: A review. Analyst. 1997;**122**(5):67R-74R

[19] Rao C, Reddi G. Platinum group metals (PGM); occurrence, use and recent trends in their determination. TrAC Trends in Analytical Chemistry. 2000;**19**(9):565-586

[20] Acres GJK, Harrison B. The development of catalysts for emission control from motor vehicles: Early research at Johnson Matthey. Topics in Catalysis. 2004;**28**(1):3-11

[21] Twigg MV. Twenty-five years of autocatalysts. Platinum Metals Review. 1999;**43**(4):168-171

[22] Wang J et al. Catalytic ozonation of dimethyl phthalate and chlorination disinfection by-product precursors over Ru/AC. Journal of Hazardous Materials. 2009;**166**(1):502-507

[23] Snyder SA, Vanderford BJ, Rexing DJ. Trace analysis of bromate, chlorate, iodate, and perchlorate in natural and bottled waters. Environmental Science & Technology. 2005;**39**(12):4586-4593

[24] Weinberg HS, Delcomyn CA, Unnam V. Bromate in chlorinated drinking waters: Occurrence and implications for future regulation. Environmental Science & Technology. 2003;**37**(14):3104-3110

[25] Butler R et al. Bromate environmental contamination: Review of impact and possible treatment. Critical Reviews in Environmental Science and Technology. 2005;**35**(3):193-217

[26] Oh BS et al. Formation of hazardous inorganic by-products during electrolysis of seawater as a disinfection process for desalination. Science of the Total Environment. 2010;**408**(23):5958-5965

[27] Boorman GA. Drinking water disinfection byproducts: Review and approach to toxicity evaluation. Environmental Health Perspectives. 1999;**107**(suppl 1):207-217

[28] McAdam E, Judd S. Biological treatment of ion-exchange brine regenerant for re-use: A review. Separation and Purification Technology. 2008;**62**(2):264-272

[29] Liu J et al. Application of a Re–Pd bimetallic catalyst for treatment of perchlorate in waste ion-exchange regenerant brine. Water Research. 2013;**47**(1):91-101

[30] Choe JK et al. Comparative assessment of the environmental sustainability of existing and emerging perchlorate treatment technologies for drinking water. Environmental Science & Technology. 2013;**47**(9):4644-4652

[31] Chaplin BP et al. Critical review of Pd-based catalytic treatment of priority contaminants in water. Environmental Science & Technology. 2012;**46**(7):3655-3670

[32] Chen X et al. Exploring beyond palladium: Catalytic reduction of aqueous oxyanion pollutants with alternative platinum group metals and new mechanistic implications. Chemical Engineering Journal. 2017;**313**:745-752

[33] Restivo J et al. Metal assessment for the catalytic reduction of bromate in water under hydrogen. Chemical Engineering Journal. 2015;**263**:119-126

[34] Soares OS et al. Bromate reduction in water promoted by metal catalysts

prepared over faujasite zeolite. Chemical Engineering Journal. 2016;**291**:199-205

[35] Chen H et al. Aqueous bromate reduction by catalytic hydrogenation over Pd/Al_2O_3 catalysts. Applied Catalysis B: Environmental. 2010;**96**(3-4):307-313

[36] Erkan A, Bakir U, Karakas G. Photocatalytic microbial inactivation over Pd doped SnO_2 and TiO_2 thin films. Journal of Photochemistry and Photobiology A: Chemistry. 2006;**184**(3):313-321

[37] Zhou B, Han S, Raja R, Somorjai G. Nanotechnology in Catalysis. New York: Kluwer Academic/Plenum Publisher; 2003

[38] Somorjai GA, Park JY. Molecular surface chemistry by metal single crystals and nanoparticles from vacuum to high pressure. Chemical Society Reviews. 2008;**37**(10):2155-2162

[39] Vries JGD. The Handbook of Homogeneous Hydrogenation. Wiley-Vch; 2007

[40] Rong Z et al. Carbon supported Pt colloid as effective catalyst for selective hydrogenation of nitroarenes to arylhydroxylamines. Chemical Communications. 2010;**46**(9):1559-1561

[41] Blaser HU, Steiner H, Studer M. Selective catalytic hydrogenation of functionalized nitroarenes: An update. ChemCatChem. 2009;**1**(2):210-221

Chapter 5

Eco-Friendly and Facile Synthesis of Substituted Imidazoles via Nano Zirconia Catalyzed One-Pot Multicomponent Reaction of Isatin Derivatives with Ammonium Acetate and Substituted Aromatic Aldehydes under Solvent Free Conditions

Sundaram Singh and Shivam Bajpai

Abstract

An eco-friendly and highly efficient approach for the synthesis of substituted imidazoles via nano zirconia catalyzed multicomponent reaction of isatin derivatives with ammonium acetate and aromatic aldehydes under solvent-free conditions has been developed. This approach can be mostly applied to medicinal chemistry due of the simple and readily available starting materials, effortless methodology, and biologically active nature of imidazoles. An additional gain of the suggested technique is the reusability of the nano ZrO_2 catalyst.

Keywords: nano ZrO_2 catalyst, multicomponent reaction, imidazole, solvent-free, Isatin

1. Introduction

Imidazole is a "1, 3-diazole" and is classified as an alkaloid. Imidazole (1) refers to the parent compound, whereas imidazoles are a class of heterocycles with similar ring structure, but varying substituents. This ring system is present in important biological building blocks, such as histidine (2), and the related hormone histamine (3). Imidazole can serve as a base and as a weak acid. Many drugs contain an imidazole ring, such as antifungal drugs and Nitroimidazole (4) [1–5].

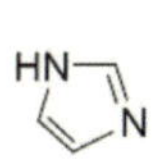

(1)

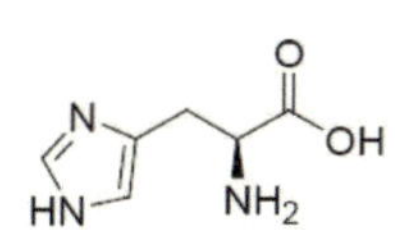

(2)

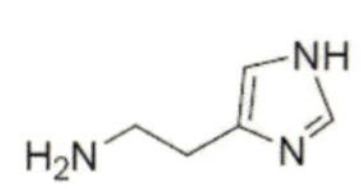

(3)

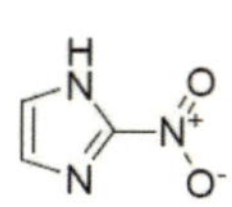

(4)

Imidazole derivatives are mostly used as organic resources [6, 7] and it also plays important roles in various types of biological activities [8, 9]. This multitalented applicability of Imidazole draws attention towards the importance of access to efficient synthetic routes to well-designed and highly substituted imidazole derivatives.

Due to their impressive significance, various synthetic routes have been designed. Substituted imidazoles are generally prepared by microwave irradiated one pot three-component cyclocondensation of a 1, 2-diketone, α-hydroxy ketone or α-ketomonoxime with an aldehyde and ammonium acetate [10–13], ionic liquids [14, 15], refluxing in acetic acid [16], silica sulfuric acid [17, 18], $Yb(OTf)_3$ [19], $Yb(OPf)_3$ [20], iodine [21], $Zr(acac)_4$ [22], $InCl_3{\cdot}3H_2O$ [23], heteropolyacid [24], sodium bisulfate [25], potassium aluminum sulfate (alum) [26], ceric ammonium nitrate (CAN) [27], $(NH_4)_6\,Mo_7O_{24}{\cdot}4H_2O$ [28], zeolite HY/silica gel [29], $ZrCl_4$ [30], polymer-supported $ZnCl_2$ [31] and L-proline [32]. Moreover, they have also been prepared by the addition of substituted amino alcohol to a thioamide and subsequent oxidation with PDC or by the reaction of aryl nitriles and α,α-dilithioarylnitromethanes or by multistep syntheses. *p*-TSA catalyzed synthesis of 2,4,5-trisubstituted imidazoles from 1,2-diketone or α-hydroxyketone, aldehyde and ammonium heptamolybdate tetrahydrate in tetrabutylammonium iodide was given by Khodaei and co-workers [33].

Due to their potential utility, majority of these synthetic routes experience one or more severe disadvantages, such as difficult and intricate work-up and purification, huge amounts of waste materials, strongly acidic conditions, occurrence of side reactions, low yields, high temperature, long reaction time and the use of expensive reagents. Hence, there is a great demand of a highly efficient protocol with mild reaction conditions to synthesize substituted imidazoles.

In recent times, metal nanoparticles are used as heterogeneous catalysts in organic synthesis mainly because they achieve the objectives of green and sustainable chemistry. Recently Scientists have done a lot of work to synthesize precise metal nanoparticles. The new path is the coherent design and synthesis of very active and selective nanocatalysts by controlling the structure and composition of the active nanoparticles among all of them. The easiness of separation, recovery, and reuse of these NPs further enhance their attractiveness as green and sustainable catalysts [3, 34–45].

Recently, nano zirconia (ZrO_2) has attracted considerable attention due to their wide applicability as a heterogeneous catalyst [46–51]. The catalytic activities and selectivities of nano zirconia are highly affected by their crystal phase (monoclinic and tetragonal) [52–54]. ZrO_2 nanoparticle catalyst is a cheap, moisture stable, safe, reusable, and commercially available white powder is of big curiosity to many researchers. It has been revealed from the literature that numerous parallel applications of nano zirconia, as an effective catalyst in green/sustainable synthetic chemistry, have already been reported [55–67].

In view of the above and as a part of our research group to synthesize the biologically active compounds [68–72], it was thought worthwhile to synthesize some novel imidazoles fused with indole nucleus of biocidal interest because the combination of two or more different heterocyclic compounds in a single molecule frequently increases the biocidal profile amazingly. With the aim of getting targeted products, i.e., substituted imidazole a greener "NOSE" (nanoparticles-catalyzed organic synthesis enhancement) approach has been designed under solvent-free conditions.

2. Results and discussion

Imidazole derivatives **4a–s** was synthesized by one pot multicomponent reaction of isatin derivatives **1a–g** with ammonium acetate **2** and substituted benzaldehydes

3a–f in the presence of catalytic amount of ZrO_2 NPs under solvent-free conditions at 110°C, in good to excellent yields (**Scheme 1**).

To optimize the reaction conditions, several parameters were tested. The catalytic efficiency of the ZrO_2 NPs was highly influenced by their amount (mol%). Therefore, a model reaction of isatin with ammonium acetate and benzaldehyde using different amounts of ZrO_2 NPs was carried out (**Table 1**). It has been observed that there is a notable impact of the catalyst on the yield of product and in the absence of catalyst only poor yield was obtained after 120 min (Entry 1, **Table 1**). It was found that product yield is increased with increasing catalyst concentration. Only 5 mol% of catalyst was required to provide 60% yield in 60 min (Entry 2, **Table 1**). The best yield of 88% was obtained with 15 mol% of ZrO_2 NPs (Entry 5, **Table 1**). However, the reaction rate and product yield were not improved by further increase of catalyst concentration (>15 mol%) (Entry 6, **Table 1**).

To optimize the molar proportion of the reactant, the model reaction was carried out using different molar proportions of reactants (**Table 2**). A scrutiny of the table clearly shows that the best result was obtained using isatin, ammonium acetate, benzaldehyde in the molar proportion 1.0:5.0:1.0 at 110°C under solvent free conditions (Entry 5, **Table 2**).

To see the effect of temperature, the model reaction was examined under different temperatures. Obviously, reaction rate and product yield both were increased with enhancing temperature from 50 to 110°C. This study shows that the 110°C was favorable temperature for the multicomponent reaction of isatin with ammonium acetate and benzaldehyde (**Table 3**).

Screening of solvent showed that solvent had a remarkable impact on the yield of product. It was observed that polar solvent provided better yield than nonpolar solvent, but excellent yield was obtained without solvent in smaller time. This may be due to the competitive adsorption of the solvent with the substrate molecule on the catalyst surface; hence reaction under solvent-free conditions gives excellent yield in short reaction time (Entry 5, **Table 4**). Another reason is that the eutectic mixture having uniform distribution of the reactants brings the reacting species in close proximity to react in solvent free condition than in the presence of solvent (Entry 1–4, **Table 4**).

Under the optimized set of reaction conditions, the effect of type of ZrO_2 (nano or bulk) was also examined using model reaction(**Table 5**). This important parameter was studied by using four concentrations 5, 10, 12 and 15 mol% of ZrO_2. These data proved that particle size and surface area would be an important factor for the catalytic efficiency of the ZrO_2 NPs.

The efficiency of the catalytic activity of the ZrO_2 NPs with several other catalysts was compared and is summarized in **Table 6**. The result indicates that ZrO_2 NPs was the best catalyst in terms of mol%, reaction time and percentage yield (**Table 6**).

Under the optimized reaction condition, the scope of this methodology was extended to the reaction of different isatin with a wide range of aromatic aldehydes. The findings reveal that the proposed methodology is equally applicable for the presence of both electron donating as well as electron withdrawing groups at the 5-position of isatin moiety (**Table 7**).

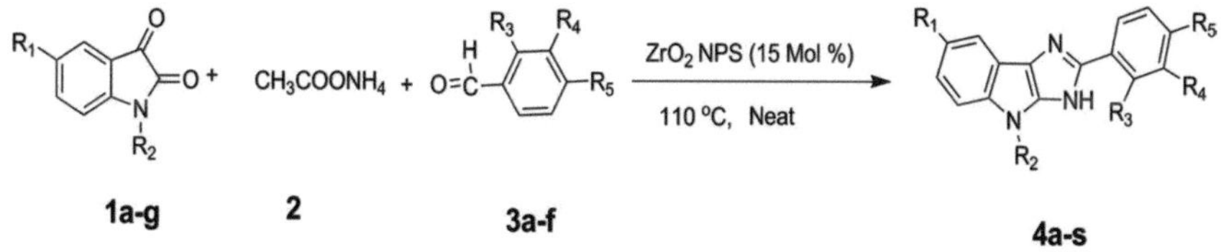

Scheme 1.
Nano ZrO_2 catalyzed synthesis of imidazole derivatives.

Entry	ZrO_2 mol%	Time (min.)	%Yield
1	0	120	23
2	5	60	60
3	10	45	75
4	12	35	82
5	**15**	**30**	**88**
6	20	30	88

Note: Bold values represent optimized reaction condition.

Table 1.
*Effect of catalyst amount (mol%) on yield of the product **4a**.*

Entry	Molar ratio of reactants	%Yield
	Isatin:ammoniumacetate:benzaldehyde	
1	1.0:1.0:1.0	Trace amount
2	1.0:2.0:1.0	35
3	1.0:3.0:1.0	52
4	1.0:4.0:1.0	78
5	**1.0:5.0:1.0**	**88**
6	1.0:6.0:1.0	87
7	1.0:5.0:1.2	87
8	1.2:5.0:1.0	86

Note: Bold values represent optimized reaction condition.

Table 2.
*Effect of molar ratio of substrates on the yield of the product **4a**.*

Entry	Temp. °C	Time	% Yield
1	rt	-	No reaction
2	50	10 h	Trace amount
3	60	6 h	65
4	70	4 h	70
5	80	1.5 h	78
6	90	55 min	84
7	100	45 min	86
8	**110**	**30 min**	**88**
9	120	30 min	88

Note: Bold values represent optimized reaction condition.

Table 3.
*Effect of temperature on the yield of the product **4a**.*

The model reaction was carried out to examine the reusability of the catalyst, After each reaction, catalyst was recovered by filtration, washed, air-dried and reused directly for the next time up to run no. 10.The results showed that there is no obvious loss in product yield in subsequent reuse which proves the reusability and recyclability of ZrO_2 NPs (**Table 8**).

Entry	Solvents	Time	%Yield
1	Ethanol	10 h	68
2	Acetonitrile	10 h	59
3	Xylene	13 h	55
4	Tolune	18 h	52
5	**Solvent free**	**30 min**	**88**

Note: Bold values represent optimized reaction condition.

Table 4.
*Effect of solvents on the yield of the product **4a**.*

Type of ZrO_2	Mol%	% Yield
ZrO_2 (Bulk)	5	42
Surface area: 6.95m^2/g	10	51
Average particle size: 2 μm	12	56
	15	66
ZrO_2 (Nano)	5	59
Surface area: 44.70 m^2/g	10	72
Average particle size: 20 nm	12	84
	15	88

[a]*Reaction condition: Isatin, ammonium acetate & benzaldehyde (1.0, 5.0, 1.0) were stirred at 110°C to produce solid product.*

Table 5.
*Effect of type of ZrO_2 (bulk & nano) on the yield of the product **4a**.*

Type of catalyst	Mol%	Time (min.)	% Yield
Bentonite clay	20	60	55
K-10 clay	20	60	58
PTSA	40	75	45
NH_4Cl	30	75	44
EDTA	40	75	40
Iodine	30	60	53
$Yb(OTf)_3$	25	60	51
TiO_2 (Nano)	20	30	80
ZrO_2 (Nano)	**15**	**30**	**88**

Note: Bold values represent optimized reaction condition.

Table 6.
*Effect of different catalysts on the yield of the product **4a**.*

The following mechanism was proposed for the formation of substituted imidazoles catalyzed by the ZrO_2 NPs is given in **Scheme 2**. The reaction proceeds via the diamine intermediate [X]. Condensation of diamine with isatin derivatives followed by dehydration, and then rearrangement through the imino intermediate [Y] yielded the desired product.

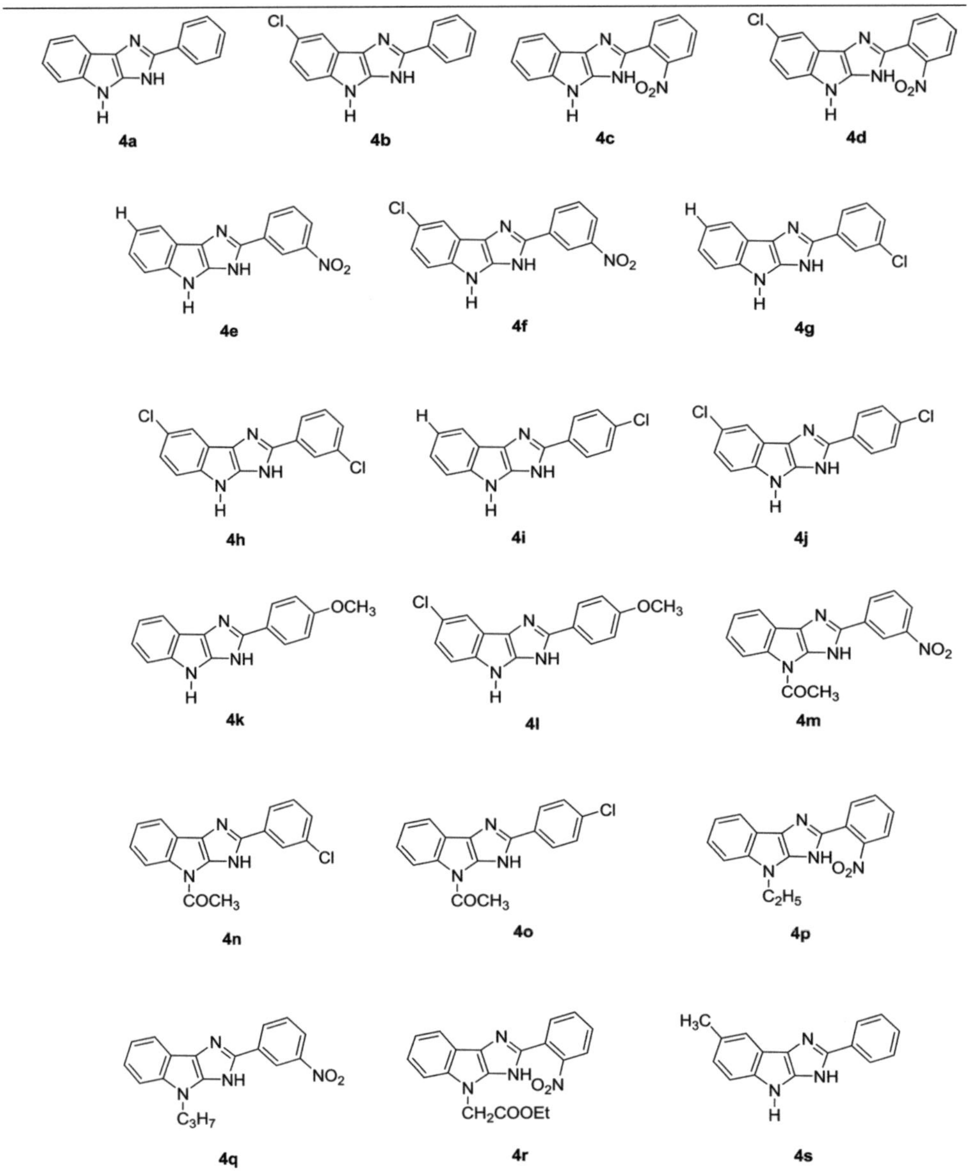

Table 7.
Synthesis of substituted imidazoles (4a–s).

ZrO_2 NPs were prepared and characterized by FTIR, XRD, SEM and TEM analysis. The BET surface area analyzer was used to calculate the specific surface area of synthesized ZrO_2 NPs.

The molecular nature of the synthesized material was identified by the FT-IR spectrum of the ZrO_2 sample. The FT-IR spectrum of ZrO_2 NPs depends on the nature of the material, preparative procedures used, solid-state structure, and so forth. In FT-IR, a strong absorption peak at about 500 cm^{-1} region is due to the Zr–O vibration, which confirm the formation of ZrO_2 structure while the peak at 751 cm^{-1} is due to stretching vibrations of Zr–O–Zr, prominent peak at 1340 cm^{-1} represents O–H bonding, peak at 1622 cm^{-1} perhaps owing to the adsorbed moisture and peaks at about 2855–2922 cm^{-1} region is due to stretching of O–H groups.

Entry	Number of cycle	%Yield
1	-	88
2	1	88[b]
3	2	87[b]
4	3	86[b]
5	4	83[b]
6	5	80[b]
7	6	80[b, c]
8	7	80[b]
9	8	78[b]
10	9	75[b]
11	10	76[b, c]

[a]*Reaction condition: Isatin, ammonium acetate, benzaldehydes (1.0:5.0:1.0) and ZrO_2 NPs (15 mol%) were stirred at 110°C to produce solid product.*

[b]*The catalyst was washed and dried at 80–90°C for 12 h.*

[c]*ZrO_2 NPs were calcinated at 600°C for 3 h.*

Table 8.
Reusability and recyclability of ZrO_2 NPs catalyst[a]*.*

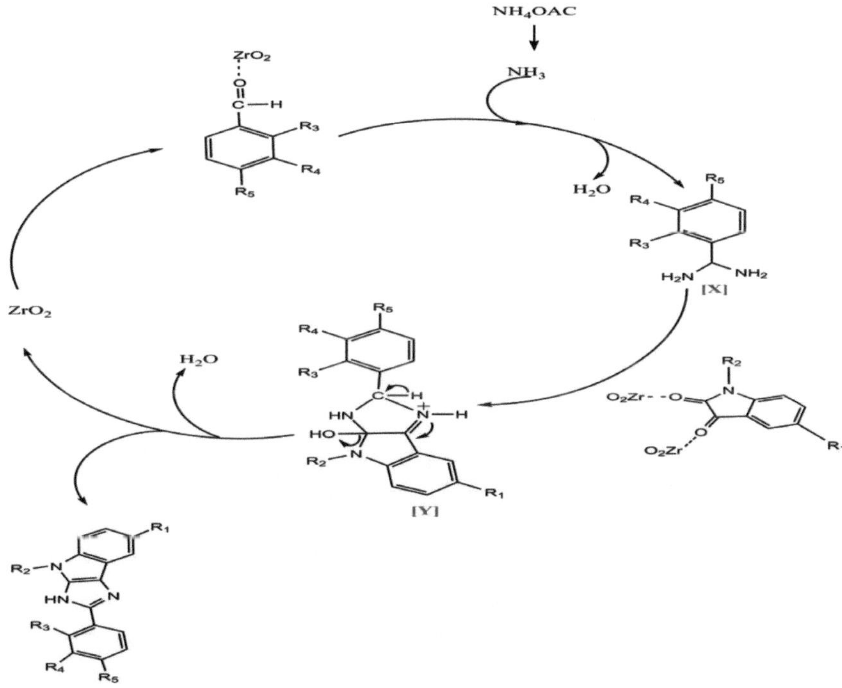

Scheme 2.
*Proposed mechanism for the formation of substituted imidazoles **4a–s**.*

The broad peaks with high intensity in XRD pattern of ZrO_2 NPs indicates that the sample was highly crystalline. The peaks observed at 2θ = 24.2 (011), 28.2 (−111), 31.4 (111), 35.0 (020), 40.5 (−112), 45.0 (211), and 55.4 (−311) are characteristics peaks of monoclinic zirconia (JCPDS card no. 37–1484) while diffraction peak observed at 2θ = 30.3 (101), 50.3 (212) and 60.2 (211) are due to tetragonal

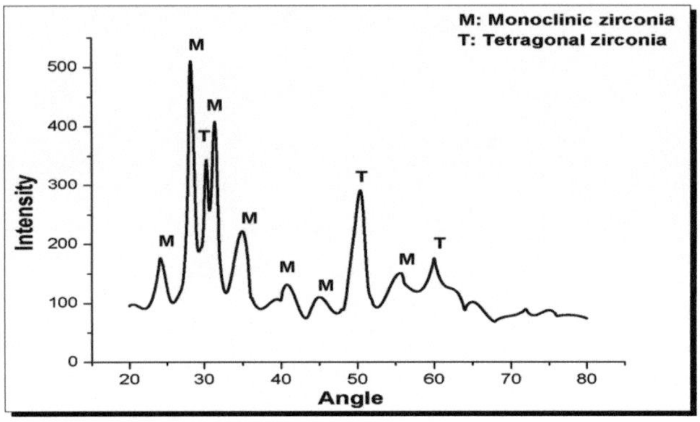

Figure 1.
XRD spectra of ZrO_2 NPs.

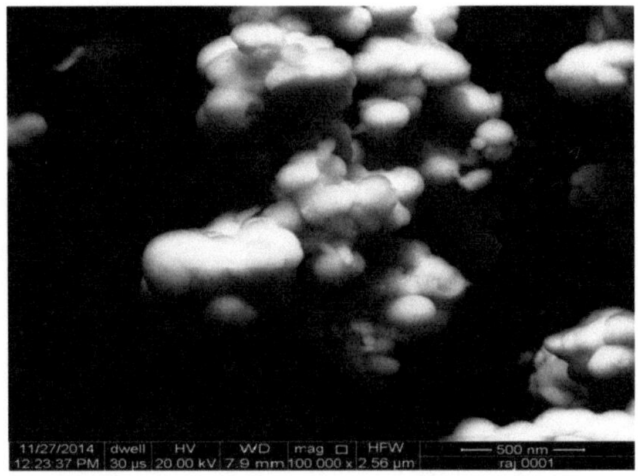

Figure 2.
SEM image of ZrO_2 NPs.

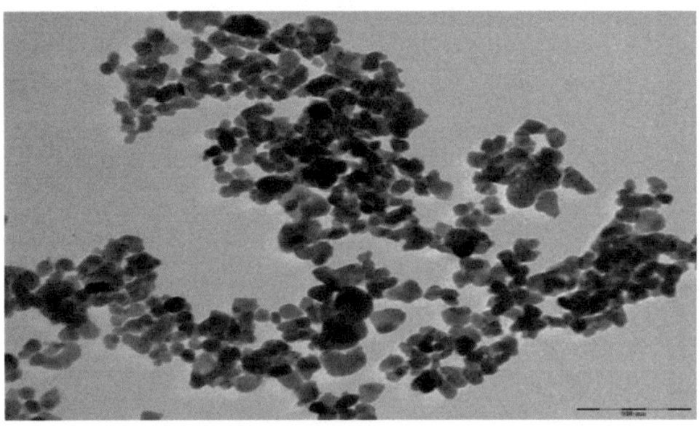

Figure 3.
TEM image of ZrO_2 NPs.

zirconia (JCPDS card no. 79-1769). The broadening of peaks shows the smaller particle size of ZrO_2 NPs **(Figure 1)**.

Morphological studies were done with the help of SEM and TEM analysis of 600°C calcinated ZrO_2 NPs sample that are shown in **Figures 2** and **3**, respectively. SEM analysis shows that NPs are non-homogenous and agglomerated and it also indicates the spherical nature and nano size (nm regime) of the ZrO_2 nanoparticles but size could be finely decided from TEM. For the purpose, TEM of sample has been shown in **Figure 3**.

As it is clear from TEM micrograph of sample, some agglomeration of the NPs, has been seen due to presence of different m- and t-phases in the sample. It was also observed that the sizes of the particles are of the order 20 nm along with agglomeration.

BET surface area analyzer was used to calculate the surface area of synthesized ZrO_2 NPs by nitrogen absorption which was found to be 44. 70 m^2/g.

3. Experimental

3.1 Typical procedure for the synthesis of ZrO_2 NPs

0.075 M solution of $ZrOCl_2.8H_2O$ was prepared and then precipitated with NH_4OH (25%) with continuous stirring on a magnetic stirrer till the P^H raises in the range of 10–10.5.This resulted in the formation of precipitate of zirconium hydroxide. The precipitate was filtered and washed with double distilled water until traces of chloride ion were completely removed from the filtrate. Complete removal of chloride ion from filtrate was checked by titrating it with $AgNO_3$ solution using potassium chromate as indicator. Now, the precipitate was dried in oven at 80–90°C for 24 h and calcinated at 600°C for 3 h in order to formation of white nano zirconia powder.

3.2 General procedure for the synthesis of substituted imidazoles 4a-s

To a mixture of isatin derivatives **1a–g** (1 mmol), ammonium acetate **2** (5 mmol), substituted aromatic aldehydes **3a–f** (1 mmol), 15 mol% of ZrO_2 NPs was added (**Scheme 1**). The mixture was heated and stirred at 110°C for 30 min. The progress of the reaction was monitored by thin layered chromatography (n-hexane:ethyl acetate, 1:1). After completion, 20 ml acetone was added to the reaction mixture; the catalyst was removed by filtration and washed with xylene and acetone. Then, 50 ml of double distilled water is added to the liquid portion. This resulted in the formation of precipitate of products **4a–s**. The precipitate was filtered, dried and recrystallized with ethanol.

3.2.1 2-Phenyl-3,4-dihydroimidazo[4,5-b]indole (4a)

Brownsolid, **IR (KBr) υ:** 3400, 3209, 3019, 2964, 1660, 1614, 1567, 1484, 1316, 1210, 1171, 1010, 877, 742, 653, 580 cm^{-1}. **1H NMR (300 MHz, DMSO) δ:** 7.80–8.86 (m, 9H, aromatic protons), 9.15 (s, 1H, NH), 9.66 (s, 1H, NH) ppm. **13C NMR (75.45 MHz, DMSO) δ:** 124.0, 126.7, 127.5, 130.2, 130.7, 132.0, 133.7, 135.5, 139.1, 148.2, 160.9 ppm. Anal. Calcd for $C_{15}H_{11}N_3$: C, 77.24; H, 4.74; N, 18.01 Found C, 77.20; H, 4.76; N, 18.03.

3.2.2 7-Chloro-2-phenyl-3,4-dihydroimidazo[4,5-b]indole (4b)

Brownsolid, **IR (KBr) υ:** 3364, 3190, 2981, 2964, 1648, 1609, 1559, 1447, 1311, 1199, 1143, 1019, 872, 744, 651, 566 cm^{-1}. **1H NMR (300 MHz, CDCl3) δ:** 7.51–8.59 (m, 8H, aromatic protons), 9.14 (s, 1H, NH), 9.45 (s, 1H, NH) ppm. **13C NMR (75.45 MHz, CDCl3) δ:** 123.9, 125.7, 128.5, 128.6, 130.3, 130.8, 132.7, 135.0, 137.5, 149.2, 159.4 ppm. Anal. Calcd for $C_{15}H_{10}ClN_3$: C, 67.28; H, 3.78; N, 15.72. Found C, 67.32; H, 3.76; N, 15.70.

3.2.3 2-(2-Nitrophenyl)-3,4-dihydroimidazo[4,5-b]indole (4c)

Brownsolid, **IR (KBr) υ:** 3332, 3201, 2995, 2917, 1658, 1623, 1549, 1485, 1348, 1280, 1176, 1068, 864, 708, 667, 544 cm^{-1}. **1H NMR (300 MHz, DMSO) δ:** 7.61–8.69 (m, 8H, aromatic protons), 9.67 (s, 1H, NH), 9.94 (s, 1H, NH) ppm. **13C NMR (75.45 MHz, DMSO) δ:** 123.8, 126.7, 128.8, 129.8, 130.1, 131.8, 135.4, 135.8, 135.9, 148.3, 160.7 ppm. Anal. Calcd for $C_{15}H_{10}N_4O_2$: C, 64.74; H, 3.62; N, 20.13 Found C, 64.69; H, 3.65; N, 20.14.

3.2.4 7-Chloro-2-(2-nitrophenyl)-3,4-dihydroimidazo[4,5-b]indole (4d)

Brownsolid, **IR (KBr) υ:** 3399, 3229, 2916, 2885, 1645, 1600, 1539, 1457, 1329, 1253, 1162, 1027, 885, 703, 647, 553 cm^{-1}. **1H NMR (300 MHz, CDCl3) δ:** 7.66–8.36 (m, 7H, aromatic protons), 8.96 (s, 1H, NH), 9.50 (s, 1H, NH) ppm. **13C NMR (75.45 MHz, CDCl3) δ:** 123.4, 123.8, 124.8, 127.1, 128.0, 128.6, 129.4, 133.9, 134.1, 134.4, 139.7, 148.7, 150.4, 160.6 ppm. Anal. Calcd for $C_{15}H_9ClN_4O_2$: C, 57.60; H, 2.91; N, 17.91 Found C, 57.51; H, 3.0; N, 17.94.

3.2.5 2-(3-Nitrophenyl)-3,4-dihydroimidazo[4,5-b]indole (4e)

Brownsolid, **IR (KBr) υ:** 3315, 3194, 3066, 2978, 1662, 1623, 1572, 1482, 1353, 1286, 1135, 1025, 832, 797, 661, 542 cm^{-1}. **1H NMR (300 MHz, DMSO) δ:** 7.58–8.55 (m, 8H, aromatic protons), 8.97 (s, 1H, NH), 9.67 (s, 1H, NH) ppm. **13C NMR (75.45 MHz, DMSO) δ:** 122.3, 123.6, 125.2, 127.9, 128.4, 129.7, 130.4, 134.4, 134.5, 135.2, 139.0, 147.6, 161.7 ppm. Anal. Calcd for $C_{15}H_{10}N_4O_2$: C, 64.70; H, 3.63; N, 20.16 Found C, 64.51; H, 3.72; N, 20.23.

3.2.6 7-Chloro-2-(3-nitrophenyl)-3,4-dihydroimidazo[4,5-b]indole (4f)

Brownsolid, **IR (KBr) υ:** 3385, 3211, 3003, 2959, 1646, 1603, 1538, 1458, 1367, 1248, 1122, 1022, 831, 741, 635, 564 cm^{-1}. **1H NMR (300 MHz, DMSO) δ:** 7.77–8.83 (m, 7H, aromatic protons), 9.13 (s, 1H, NH), 9.62 (s, 1H, NH) ppm. **13C NMR (75.45 MHz, DMSO) δ:** 124.0, 126.6, 127.5, 127.6, 130.2, 130.7, 132.0, 133.6, 135.4, 139.1, 148.2, 160.8, 160.9 ppm. Anal. Calcd for $C_{15}H_9ClN_4O_2$: C, 57.61; H, 2.90; N, 17.92 Found C, 57.67; H, 2.90; N, 17.90.

3.2.7 2-(3-Chlorophenyl)-3,4-dihydroimidazo[4,5-b]indole (4g)

Brownsolid, **IR (KBr) υ:** 3405, 3217, 2948, 2909, 1671, 1617, 1568, 1454, 1371, 1283, 1134, 1018, 892, 754, 641, 577 cm^{-1}. **1H NMR (300 MHz, DMSO) δ:** 7.60–8.56 (m, 9H, aromatic protons and 1H, NH), 9.69 (s, 1H, NH) ppm. **13C NMR (75.45 MHz, DMSO) δ:** 123.8, 126.6, 128.8, 128.7, 130.1, 131.0, 131.6, 135.2, 137.0, 148.4, 160.6 ppm. Anal. Calcd for $C_{15}H_{10}ClN_3$: C, 67.30; H, 3.77; N, 15.70 Found C, 67.29; H, 3.75; N, 15.70.

3.2.8 7-Chloro-2-(3-chlorophenyl)-3,4-dihydroimidazo[4,5-b]indole (4h)

Brownsolid, **IR (KBr) υ:** 3398, 3227, 2977, 2893, 1664, 1605, 1551, 1477, 1358, 1242, 1163, 1011, 844, 743, 650, 567 cm^{-1}. **1H NMR (300 MHz, CDCl3) δ:** 7.62–8.31 (m, 7H, aromatic protons), 8.90 (s, 1H, NH), 9.42 (s, 1H, NH) ppm. **13C NMR (75.45 MHz, CDCl3) δ:** 124.1, 125.8, 126.6, 128.6, 129.8, 130.3, 130.7, 133.2, 134.8, 135.2, 139.3, 149.0, 160.5 ppm. Anal. Calcd for $C_{15}H_9C_{l2}N_3$: C, 59.62; H, 3.00; N, 13.91 Found C, 59.52; H, 3.05; N, 13.89.

3.2.9 2-(4-Chlorophenyl)-3,4-dihydroimidazo[4,5-b]indole (4i)

Brownsolid, **IR (KBr) υ:** 3362, 3255, 3015, 2882, 1669, 1620, 1565, 1482, 1375, 1235, 1140, 1026, 890, 777, 663, 526 cm^{-1}. **1H NMR (300 MHz, CDCl3) δ:** 7.48–8.58 (m, 8H, aromatic protons), 9.45 (s, 1H, NH), 10.16 (s, 1H, NH) ppm. **13C NMR (75.45 MHz, CDCl3) δ:** 123.3, 127.8, 127.9, 128.7, 129.7, 134.9, 135.6, 136.2, 149.7, 161.3 ppm. Anal. Calcd for $C_{15}H_{10}ClN_3$: C, 67.30; H, 3.77; N, 15.70 Found C, 67.31; H, 3.75; N, 15.73.

3.2.10 7-Chloro-2-(4-chlorophenyl)-3,4-dihydroimidazo[4,5-b]indole (4j)

Brownsolid, **IR (KBr) υ:** 3386, 3233, 3047, 2960, 1657, 1612, 1558, 1435, 1348, 1282, 1153, 1019, 871, 742, 654, 552 cm^{-1}. **1H NMR (300 MHz, CDCl3) δ:** 7.46–8.54 (m, 7H, aromatic protons), 8.99 (s, 1H, NH), 9.35 (s, 1H, NH) ppm. **13C NMR (75.45 MHz, CDCl3) δ:** 123.9, 125.8, 128.8, 129.8, 130.3, 133.0, 135.2, 136.0, 137.1, 149.1, 159.5 ppm. Anal. Calcd for $C_{15}H_9Cl_2N_3$: C, 59.62; H, 3.00; N, 13.91 Found C, 59.55; H, 3.10; N, 13.90.

3.2.11 2-(4-Methoxyphenyl)-3,4-dihydroimidazo[4,5-b]indole (4k)

Brownsolid, **IR (KBr) υ:** 3351, 3138, 3001, 2944, 2881, 1667, 1619, 1575, 1450, 1371, 1284, 1157, 1021, 863, 743, 654, 534 cm^{-1}. **1H NMR (300 MHz, DMSO) δ:** 4.00 (s, 3H, CH_3), 7.26–8.69 (m, 9H, aromatic protons and 1H, NH), 9.72 (s, 1H, NH) ppm. **13C NMR (75.45 MHz, DMSO) δ:** 56.9, 122.2, 123.2, 125.9, 127.7, 127.4, 127.9, 128.7, 129.0, 129.6, 130.7, 131.2, 131.4, 131.5, 138.5, 139.7, 140.1, 143.9, 145.8, 154.6 ppm. Anal. Calcd for $C_{16}H_{13}N_3O$: C, 72.99; H, 4.98; N, 15.96 Found C, 72.91; H, 5.04; N, 15.95.

3.2.12 7-Chloro-2-(4-methoxyphenyl)-3,4-dihydroimidazo[4,5-b]indole (4l)

Brownsolid, **IR (KBr) υ:** 3370, 3259, 2991, 2911, 1675, 1614, 1558, 1480, 1436 1377, 1291, 1186, 1049, 869, 745, 651, 522 cm^{-1}. **1H NMR (300 MHz, CDCl3) δ:** 3.61 (s, 3H, CH_3), 7.44–8.47 (m, 8H, aromatic protons), 8.58 (s, 1H, NH), 9.36 (s, 1H, NH) ppm. **13C NMR (75.45 MHz, CDCl3) δ:** 61.8, 122.16, 122.29, 124.0, 125.3, 126.6, 130.1, 130.4, 132.3, 133.9, 135.5, 138.5, 148.1, 148.2, 157.8 ppm. Anal. Calcd for $C_{16}H_{12}ClN_3O$: C, 64.54; H, 4.06; N, 14.11 Found C, 64.70; H, 4.00; N, 14.10.

3.2.13 1-(2-(3-Nitrophenyl)imidazo[4,5-b]indol-4(3H)-yl)ethanone (4m)

Brownsolid, **IR (KBr) υ:** 3389, 3266, 2978, 2935, 1694, 1645, 1616, 1571, 1467, 1346, 1224, 1133, 1021, 823, 744, 641, 572 cm^{-1}. **1H NMR (300 MHz, CDCl3) δ:** 1.91 (s, 3H, CH_3), 7.22–8.51 (m, 8H, aromatic protons), 9.40 (s, 1H, NH) ppm. **13C NMR (75.45 MHz, CDCl3) δ:** 23.7, 122.1, 122.4, 123.2, 123.5, 124.8, 127.2, 127.4, 127.6, 128.2, 129.7, 130.1, 130.9, 133.0, 134.1, 134.4, 135.6, 138.0, 141.0, 148.7, 150.4, 160.3, 172.2 ppm. Anal. Calcd for $C_{17}H_{12}N_4O_3$: C, 63.75; H, 3.78; N, 17.49 Found C, 63.68; H, 3.88; N, 17.52.

3.2.14 1-(2-(3-Chlorophenyl)imidazo[4,5-b]indol-4(3H)-yl)ethanone (4n)

Brownsolid, **IR (KBr) υ:** 3367, 3215, 2947, 2923, 1692, 1662 1607, 1580, 1477, 1359, 1272, 1144, 1042, 807, 735, 653, 546 cm^{-1}. **1H NMR (300 MHz, CDCl3) δ:** 1.25 (s, 3H, CH_3), 7.34–8.11 (m, 8H, aromatic protons), 9.36 (s, 1H, NH) ppm. **13C NMR (75.45 MHz, CDCl3) δ:** 23.6, 122.5, 124.1, 125.3, 126.6, 130.1, 130.4, 130.7,

132.3, 133.9, 134.6, 135.5, 138.5, 148.1, 148.2, 157.8, 160.9, 166.2 ppm. Anal. Calcd for $C_{17}H_{12}ClN_3O$: C, 65.92; H, 3.90; N, 13.57 Found C, 66.01; H, 3.95; N, 13.47.

3.2.15 1-(2-(4-Chlorophenyl)imidazo[4,5-b]indol-4(3H)-yl)ethanone (4o)

Brownsolid, **IR (KBr) υ:** 3350, 3285, 3011, 2935, 1685, 1654 1611, 1572, 1485, 1455, 1343, 1284, 1132, 1062, 899, 783, 659, 531 cm^{-1}. **1H NMR (300 MHz, CDCl3) δ:** 1.66 (s, 3H, CH_3), 7.35–8.12 (m, 8H, aromatic protons), 9.37 (s, 1H, NH) ppm. **13C NMR (75.45 MHz, CDCl3) δ:** 23.5, 123.3, 125.6, 127.2, 128.1, 128.4, 128.7, 129.0, 129.2, 129.5, 133.1, 134.3, 134.6, 134.7, 136.4, 137.0, 142.0, 150.5, 160.2, 160.9 ppm. Anal. Calcd for $C_{17}H_{12}ClN_3O$: C, 65.92; H, 3.90; N, 13.57 Found C, 65.99; H, 3.93; N, 13.54.

3.2.16 4-Ethyl-2-(2-nitrophenyl)-3,4-dihydroimidazo[4,5-b]indole (4p)

Brownsolid, **IR (KBr) υ:** 3416, 3199, 3012, 2999, 2942, 2872, 1654, 1607, 1561, 1441, 1453 1351, 1283, 1192, 1021, 861, 741, 657, 526 cm^{-1}. **1H NMR (300 MHz, DMSO) δ:** 1.41–1.45 (t, J = 6.6, 3H, CH_3), 4.38–4.45 (q, J = 6.9, 2H, CH_2), 7.53–8.63 (m, 8H, aromatic protons), 9.63 (s, 1H, NH) ppm. **13C NMR (75.45 MHz, DMSO) δ:** 12.5, 24.9, 122.1, 122.8, 123.1, 124.6, 127.5, 127.8, 130.2, 130.4, 130.7, 133.5, 134.4, 137.7, 135.4, 136.4, 137.5, 140.7, 148.3, 149.7, 159.9 ppm. Anal. Calcd for $C_{17}H_{14}N_4O_2$: C, 66.66; H, 4.61; N, 18.29 Found C, 66.74; H, 4.65; N, 18.20.

3.2.17 2-(3-Nitrophenyl)-4-propyl-3,4-dihydroimidazo[4,5-b]indole (4q)

Brownsolid, **IR (KBr) υ:** 3400, 3301, 3221, 3135, 3009, 2951, 2912, 2865, 1664, 1616, 1571, 1478, 1422, 1371, 1271, 1181, 1037, 873, 739, 649, 536 cm^{-1}. **1H NMR (300 MHz, CDCl3) δ:** 1.05–1.10 (t, J = 6.9, 3H, CH_3), 1.70–1.82 (m, 2H, CH_2), 3.19–3.24 (t, J = 6.6, 2H, CH_2), 7.47–8.41 (m, 8H, aromatic protons), 8.94 (s, 1H, NH) ppm. **13C NMR (75.45 MHz, CDCl3) δ:** 10.1, 21.3, 44.0, 121.5, 124.2, 125.2, 125.5, 125.6, 129.3, 136.1, 136.7, 140.5, 150.9, 159.3 ppm. Anal. Calcd for $C_{18}H_{16}N_4O_2$ C, 67.49; H, 5.03; N, 17.49 Found C, 67.44; H, 5.10; N, 17.52.

3.2.18 Ethyl 2-(2-(2-nitrophenyl)imidazo[4,5-b]indol-4(3H)-yl)acetate (4r)

Brownsolid, **IR (KBr) υ:** 3389, 3255, 3129, 3116, 3027, 2969, 2913, 2847, 1735, 1657, 1618, 1569, 1435, 1353, 1264, 1158, 1049, 854, 751, 651, 546 cm^{-1}. **1H NMR (300 MHz, DMSO) δ:** 1.22–1.27 (t, J = 7.2, 3H, CH_3), 4.19–4.26 (q, J = 6.9, 2H, CH_2), 5.32 (s, 2H, CH_2), 7.37–8.28 (m, 8H, aromatic protons), 9.63 (s, 1H, NH) ppm. **13C NMR (75.45 MHz, DMSO) δ:** 15.0, 52.6, 65.1, 122.3, 123.6, 125.2, 127.9, 128.4, 130.5, 134.0, 135.2, 139.0, 148.3, 149.6, 161.7, 171.0 ppm. Anal. Calcd for $C_{19}H_{16}N_4O_4$: C, 62.63; H, 4.43; N, 15.38 Found C, 62.71; H, 4.51; N, 15.30.

3.2.19 7-Methyl-2-phenyl-3,4-dihydroimidazo[4,5-b]indole (4 s)

Brownish whitesolid, **IR (KBr) υ:** 3398, 3242, 2963, 2931, 1648, 1607, 1559, 1451, 1311, 1232, 1142, 1027, 813, 741, 655, 534 cm^{-1}. **1H NMR (300 MHz, DMSO) δ:** 2.22 (s, 3H, CH_3), 7.55–8.57 (m, 8H, aromatic protons), 9.69 (s, 1H, NH), 10.10 (s, 1H, NH) ppm. **13C NMR (75.45 MHz, DMSO) δ:** 23.9, 122.9, 123.1, 127.7, 128.5, 129.6, 130.2, 130.7, 131.3, 133.3, 136.8, 146.3, 154.1, 161.2 ppm. Anal. Calcd for $C_{16}H_{13}N_3$: C, 77.71; H, 5.30; N, 16.99 Found C, 77.64; H, 5.34; N, 17.02.

4. Conclusion

ZrO_2 nanoparticles have been synthesized and a novel synthetic route has been developed for the multicomponent reaction of isatin derivatives with ammonium acetate and substituted aromatic aldehydes using ZrO_2 nanoparticles under solvent-free conditions. The yields of the products obtained were up to 93% at 110°C. The advantage of the proposed method is its facile reaction conditions; the product can be isolated very easily without the use of column chromatography and the catalyst can be recycled. The simplicity of the presented protocol makes it an interesting alternative to other approaches. The obtained catalyst is expected to contribute to the development of environmentally benign methods and forms a part of nanomaterial chemistry.

A. Appendix

NMR spectra of compound **4a**

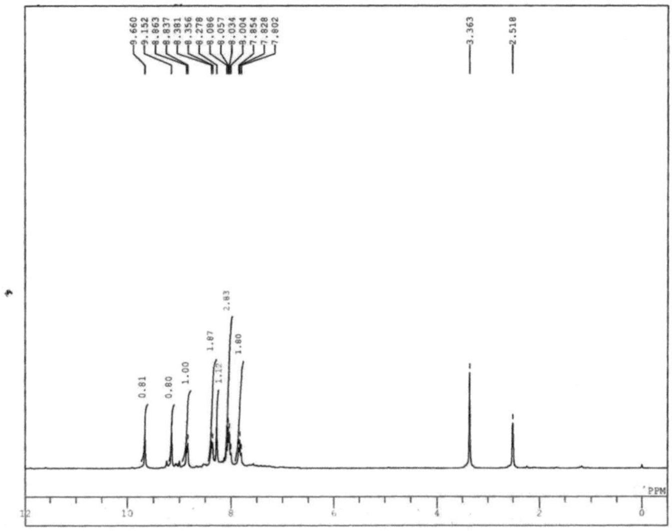

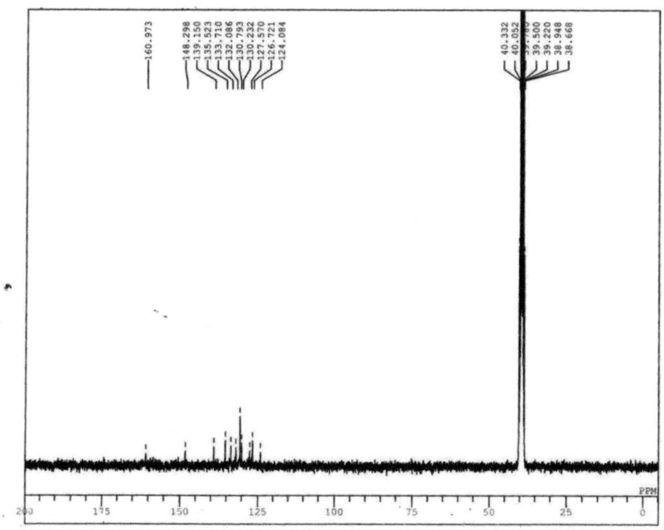

Author details

Sundaram Singh* and Shivam Bajpai
Indian Institute of Technology (BHU), Varanasi, India

*Address all correspondence to: sundaram.apc@itbhu.ac.in

References

[1] Grimmett M, Katritzky A. In: Rees CW, Katritzky AR, Rees CW, editors. Comprehensive Heterocyclic Chemistry. Vol. 5. New York: Pergamon Press; 1984. pp. 345-456

[2] Grimmett MR. Imidazole and Benzimidazole Synthesis. Cambridge, Massachusetts, USA: Academic Press; 1997

[3] Grunes J, Zhu J, Somorjai GA. Catalysis and nanoscience. Chemical Communications. 2003;**18**:2257-2260

[4] Pozharskii AF, Soldatenkov AT, Katritzky AR. Heterocycles and health. In: Heterocycles in Life and Society: An Introduction to Heterocyclic Chemistry, Biochemistry and Applications. 2nd ed. 1997. pp. 139-183

[5] Brown EG. Ring Nitrogen and Key Biomolecules: The Biochemistry of N-Heterocycles. Berlin, Germany: Springer Science & Business Media; 2012

[6] Gostev F, Kol'tsova L, Petrukhin A, Titov A, Shiyonok A, Zaichenko N. Spectral luminescent properties and dynamics of intramolecular processes in 2,4,5-triarylimidazoles. Journal of Photochemistry and Photobiology, A: Chemistry. 2003;**156**:15-22

[7] Park S, Kwon O-H, Kim S, Park S, Choi M-G, Cha M, et al. Imidazole-based excited-state intramolecular proton-transfer materials: Synthesis and amplified spontaneous emission from a large single crystal. Journal of the American Chemical Society. 2005;**127**:10070-10074

[8] Kruse LI, Ladd DL, Harrsch PB, McCabe FL, Mong SM, Faucette L, et al. Synthesis, tubulin binding, antineoplastic evaluation, and structure-activity relationship of oncodazole analogs. Journal of Medicinal Chemistry. 1989;**32**:409-417

[9] Bunnage ME, Owen DR. TAFIa inhibitors in the treatment of thrombosis. Current Opinion in Drug Discovery & Development. 2008;**11**:480-486

[10] Usyatinsky AY, Khmelnitsky YL. Microwave-assisted synthesis of substituted imidazoles on a solid support under solvent-free conditions. Tetrahedron Letters. 2000;**41**:5031-5034

[11] Sparks RB, Combs AP. Microwave-assisted synthesis of 2,4,5-triaryl-imidazole; a novel thermally induced N-hydroxyimidazole NO bond cleavage. Organic Letters. 2004;**6**:2473-2475

[12] Wolkenberg SE, Wisnoski DD, Leister WH, Wang Y, Zhao Z, Lindsley CW. Efficient synthesis of imidazoles from aldehydes and 1,2-diketones using microwave irradiation. Organic Letters. 2004;**6**:1453-1456

[13] Oskooie HA, Alimohammadi Z, Heravi MM. Microwave-assisted solid-phase synthesis of 2,4,5-triaryl imidazoles in solventless system: An improved protocol. Heteroatom Chemistry. 2006;**17**:699-702

[14] Siddiqui SA, Narkhede UC, Palimkar SS, Daniel T, Lahoti RJ, Srinivasan KV. Room temperature ionic liquid promoted improved and rapid synthesis of 2,4,5-triaryl imidazoles from aryl aldehydes and 1,2 diketones or α-hydroxyketone. Tetrahedron. 2005;**61**:3539-3546

[15] Xia M, Lu Y-d. A novel neutral ionic liquid-catalyzed solvent-free synthesis of 2,4,5-trisubstituted imidazoles under microwave irradiation. Journal of Molecular Catalysis A: Chemical. 2007;**265**:205-208

[16] Chary MV, Keerthysri NC, Vupallapati SV, Lingaiah N, Kantevari S. Tetrabutylammonium

bromide (TBAB) in isopropanol: An efficient, novel, neutral and recyclable catalytic system for the synthesis of 2,4,5-trisubstituted imidazoles. Catalysis Communications. 2008;**9**:2013-2017

[17] Shaabani A, Rahmati A. Silica sulfuric acid as an efficient and recoverable catalyst for the synthesis of trisubstituted imidazoles. Journal of Molecular Catalysis A: Chemical. 2006;**249**:246-248

[18] Shaabani A, Rahmati A, Farhangi E, Badri Z. Silica sulfuric acid promoted the one-pot synthesis of trisubstituted imidazoles under conventional heating conditions or using microwave irradiation. Catalysis Communications. 2007;**8**:1149-1152

[19] Wang L-M, Wang Y-H, Tian H, Yao Y-F, Shao J-H, Liu B. Ytterbium triflate as an efficient catalyst for one-pot synthesis of substituted imidazoles through three-component condensation of benzil, aldehydes and ammonium acetate. Journal of Fluorine Chemistry. 2006;**127**:1570-1573

[20] Shen M-G, Cai C, Yi W-B. Ytterbium perfluorooctanesulfonate as an efficient and recoverable catalyst for the synthesis of trisubstituted imidazoles. Journal of Fluorine Chemistry. 2008;**129**:541-544

[21] Kidwai M, Mothsra P, Bansal V, Somvanshi RK, Ethayathulla AS, Dey S, et al. One-pot synthesis of highly substituted imidazoles using molecular iodine: A versatile catalyst. Journal of Molecular Catalysis A: Chemical. 2007;**265**:177-182

[22] Khosropour AR. Ultrasound-promoted greener synthesis of 2,4,5-trisubstituted imidazoles catalyzed by Zr (acac) 4 under ambient conditions. Ultrasonics Sonochemistry. 2008;**15**:659-664

[23] Sharma SD, Hazarika P, Konwar D. An efficient and one-pot synthesis of 2,4,5-trisubstituted and 1,2,4,5-tetrasubstituted imidazoles catalyzed by $InCl_3 \cdot 3H_2O$. Tetrahedron Letters. 2008;**49**:2216-2220

[24] Heravi MM, Sadjadi S, Oskooie HA, Hekmatshoar R, Bamoharram FF. The one-pot synthesis of 2,4,5-triaryl-imidazoles using heteropolyacids as heterogeneous and recyclable catalysts. Journal of the Chinese Chemical Society. 2008;**55**:1199-1203

[25] Sangshetti JN, Kokare ND, Kotharkar SA, Shinde DB. Sodium bisulfite as an efficient and inexpensive catalyst for the one-pot synthesis of 2,4,5-triaryl-1H-imidazoles from benzil or benzoin and aromatic aldehydes. Monatshefte für Chemie/Chemical Monthly. 2008;**139**:125-127

[26] Mohammadi AA, Mivechi M, Kefayati H. Potassium aluminum sulfate (alum): An efficient catalyst for the one-pot synthesis of trisubstituted imidazoles. Monatshefte für Chemie/Chemical Monthly. 2008;**139**:935-937

[27] Shaabani A, Maleki A, Behnam M. Tandem oxidation process using ceric ammonium nitrate: Three-component synthesis of trisubstituted imidazoles under aerobic oxidation conditions. Synthetic Communications. 2008;**39**:102-110

[28] Safari J, Khalili SD, Banitaba SH. A novel and an efficient catalyst for one-pot synthesis of 2,4,5-trisubstituted imidazoles by using microwave irradiation under solvent-free conditions. Journal of Chemical Sciences. 2010;**122**:437-441

[29] Balalaie S, Arabanian A, Hashtroudi MS. Zeolite HY and silica gel as new and efficient heterogenous catalysts for the synthesis of triarylimidazoles under microwave irradiation. Monatshefte für Chemie/Chemical Monthly. 2000;**131**:945-948

[30] Sharma G, Jyothi Y, Lakshmi PS. Efficient room-temperature synthesis of tri-and tetrasubstituted imidazoles catalyzed by $ZrCl_4$. Synthetic Communications. 2006;**36**:2991-3000

[31] Wang L, Cai C. Polymer-supported zinc chloride: A highly active and reusable heterogeneous catalyst for one-pot synthesis of 2,4,5-trisubstituted imidazoles. Monatshefte für Chemie/ Chemical Monthly. 2009;**140**:541-546

[32] Samai S, Nandi GC, Singh P, L-Proline SM. An efficient catalyst for the one-pot synthesis of 2,4,5-trisubstituted and 1,2,4,5-tetrasubstituted imidazoles. Tetrahedron. 2009;**65**:10155-10161

[33] Khodaei MM, Bahrami K, Kavianinia I. p-TSA catalyzed synthesis of 2,4,5-triarylimidazoles from ammonium heptamolybdate tetrahydrate in TBAI. Journal of the Chinese Chemical Society. 2007;**54**:829-833

[34] Reetz MT, Westermann E. Phosphane-free palladium-catalyzed coupling reactions: The decisive role of Pd nanoparticles. Angewandte Chemie, International Edition. 2000;**39**:165-168

[35] Ramarao C, Ley SV, Smith SC, Shirley IM, DeAlmeida N. Encapsulation of palladium in polyurea microcapsules. Chemical Communications. 2002:1132-1133

[36] Gladysz JA. Recoverable catalysts. Ultimate goals, criteria of evaluation, and the green chemistry interface. Pure and Applied Chemistry. 2001;**73**:1319-1324

[37] Gladysz JA. Introduction: Recoverable catalysts and reagents perspective and prospective. Chemical Reviews. 2002;**102**:3215-3216

[38] Pacchioni G. Quantum chemistry of oxide surfaces: From CO chemisorption to the identification of the structure and nature of point defects on MgO. Surface Review and Letters. 2000;**7**:277-306

[39] Cox D, Trevor D, Whetten R, Kaldor A. Aluminum clusters: Ionization thresholds and reactivity toward deuterium, water, oxygen, methanol, methane, and carbon monoxide. The Journal of Physical Chemistry. 1988;**92**:421-429

[40] Polshettiwar V, Varma RS. Green chemistry by nano-catalysis. Green Chemistry. 2010;**12**:743-754

[41] Polshettiwar V, Baruwati B, Varma RS. Self-assembly of metal oxides into three-dimensional nanostructures: Synthesis and application in catalysis. ACS Nano. 2009;**3**:28-736

[42] Polshettiwar V, Nadagouda MN, Varma RS. The synthesis and applications of a micro-pine-structured nanocatalyst. Chemical Communications. 2008:6318-6320

[43] Shimizu K, Sato R, Satsuma A. Direct C-C cross-coupling of secondary and primary alcohols catalyzed by a γ-alumina-supported silver subnanocluster. Angewandte Chemie. 2009;**121**:4042-4046

[44] Murugadoss A, Goswami P, Paul A, Chattopadhyay A. 'Green' chitosan bound silver nanoparticles for selective C-C bond formation via in situ iodination of phenols. Journal of Molecular Catalysis A: Chemical. 2009;**304**:153-158

[45] Witham CA, Huang W, Tsung C-K, Kuhn JN, Somorjai GA, Toste FD. Converting homogeneous to heterogeneous in electrophilic catalysis using monodisperse metal nanoparticles. Nature Chemistry. 2010;**2**:36-41

[46] Liu G, Lin Y. Electrochemical sensor for organophosphate

pesticides and nerve agents using zirconia nanoparticles as selective sorbents. Analytical Chemistry. 2005;**77**:5894-5901

[47] Liu S, Han MY. Silica-coated metal nanoparticles. Chemistry - An Asian Journal. 2010;**5**:36-45

[48] Lu J, Zang J, Shan S, Huang H, Wang Y. Synthesis and characterization of core– shell structural MWNT–zirconia nanocomposites. Nano Letters. 2008;**8**:4070-4074

[49] Sholklapper TZ, Radmilovic V, Jacobson CP, Isco SJ, De Jonghe LC. Synthesis and stability of a nanoparticle-infiltrated solid oxide fuel cell electrode. Electrochemical and Solid-State Letters. 2007;**10**:B74-B76

[50] Luo X, Morrin A, Killard AJ, Smyth MR. Application of nanoparticles in electrochemical sensors and biosensors. Electroanalysis. 2006;**18**:319-326

[51] Steiner SA, Baumann TF, Bayer BC, Blume R, Worsley MA, MoberlyChan WJ, et al. Nanoscale zirconia as a nonmetallic catalyst for graphitization of carbon and growth of single-and multiwall carbon nanotubes. Journal of the American Chemical Society. 2009;**131**:12144-12154

[52] Rhodes MD, Bell AT. The effects of zirconia morphology on methanol synthesis from CO and H_2 over Cu/ZrO_2 catalysts: Part I. Steady-state studies. Journal of Catalysis. 2005;**233**:198-209

[53] Stichert W, Schüth F, Kuba S, Knözinger H. Monoclinic and tetragonal high surface area sulfated zirconias in butane isomerization: CO adsorption and catalytic results. Journal of Catalysis. 2001;**198**:277-285

[54] Li W, Yin YQ, Gao RX, Hou RL. Structures and properties of zirconia-supported ruthenium oxide catalysts for the selective oxidation of methanol to methyl formate. Journal of Molecular Catalysis. 1999;**13**:186-192

[55] Yamaguchi T, Tanabe K, Kung YC. Preparation and characterization of ZrO_2 and SO_4^{2-}-promoted ZrO_2. Materials Chemistry and Physics. 1987;**16**:67-77

[56] Yamaguchi T. Application of ZrO_2 as a catalyst and a catalyst support. Catalysis Today. 1994;**20**:199-217

[57] Asakura K, Aoki M, Iwasawa Y. Selective isopentane formation from CH_3OH on a new one-atomic layer ZrO2/ZSM-5 hybrid catalyst. Catalysis Letters. 1988;**1**:395-403

[58] Li W, Liu H, Iglesia E. Structures and properties of zirconia-supported ruthenium oxide catalysts for the selective oxidation of methanol to methyl formate. The Journal of Physical Chemistry B. 2006;**110**:23337-23342

[59] He D, Ding Y, Luo H, Li C. Effects of zirconia phase on the synthesis of higher alcohols over zirconia and modified zirconia. Journal of Molecular Catalysis A: Chemical. 2004;**208**:267-271

[60] Mercera PDL, Ommen VJG, Doesburg EBM, Burggraat AJ, Ross JRH. Zirconia as a support for catalysts: Evolution of the texture and structure on calcination in air. Applied Catalysis. 1990;**57**:127-148

[61] Mercera PDL, Ommen VJG, Doesburg EBM, Burggraat AJ, Ross JRH. Zirconia as a support for catalysts: Influence of additives on the thermal stability of the porous texture of monoclinic zirconia. Applied Catalysis. 1991;**71**:363-391. DOI: 10.1016/0166-9834(91)85092-A

[62] Damyanova S, Grange P, Delmon B. Surface characterization of zirconia coated alumina and silica carriers. Journal of Catalysis. 1997;**168**:421-430

[63] Khodakov A, Olthof B, Bell AT, Iglesia E. Ultrasound-promoted greener synthesis of 2,4,5-trisubstituted imidazoles catalyzed by Zr (acac) 4 under ambient conditions. Journal of Catalysis. 1999;**181**:205-216

[64] Tsipouriari V, Efstathiou A, Zhang Z, Verykios X. Reforming of methane with carbon dioxide to synthesis gas over supported Rh catalysts. Catalysis Today. 1994;**21**:579-587

[65] Maity S, Rana M, Srinivas B, Bej S, Dhar GM, Rao TP. Characterization and evaluation of ZrO_2 supported hydrotreating catalysts. Journal of Molecular Catalysis A: Chemical. 2000;**153**:121-127

[66] Xie S, Iglesia E, Bell AT. Water-assisted tetragonal-to-monoclinic phase transformation of ZrO_2 at low temperatures. Chemistry of Materials. 2000;**12**:2442-2447

[67] Chuah G, Liu S, Jaenicke S, Li J. High surface area zirconia by digestion of zirconium propoxide at different pH. Microporous and Mesoporous Materials. 2000;**39**:381-392

[68] Bajpai S, Singh S, Srivastava V. An easy and efficient protocol for the condensation reaction of isatin and N-substituted isatins with 1,2-diaminobenzene using low cost reusable clay catalyst. Heterocyclic Communications. 2014;**20**:37-39

[69] Bajpai S, Singh S, Srivastava V. Rutile phase nanoTiO2 as an effective heterogeneous catalyst for condensation reaction of isatin derivatives with 1,2-diaminobenzene under solvent free conditions: A greener "NOSE" approach. Arabian Journal of Chemistry. 2014. DOI: 10.1016/j.arabjc.2014.11.037

[70] Bajpai S, Singh S, Srivastava V. Monoclinic zirconia nanoparticle-catalyzed regioselective synthesis of some novel substituted spirooxindoles through one-pot multicomponent reaction in a ball mill: A step toward green and sustainable chemistry. Synthetic Communications. 2017;**47**:1514-1525

[71] Gajaganti S, Bajpai S, Srivastava V, Singh S. An efficient room temperature oxygen radical anion($O^{2\bullet-}$) mediated one-pot multi-component synthesis of spirooxindoles. Canadian Journal of Chemistry. 2017;**95**:1296-1302

[72] Mishra A, Singh S, Srivastava V. Cerium catalyzed transamidation of secondary amides under ultrasound irradiation: A breakthrough in organic synthesis. Asian Journal of Organic Chemistry. 2018;**7**:1600-1604

Chapter 6

Transition Metal Chalcogenide (TMC) Nanocomposites for Environmental Remediation Application over Extended Solar Irradiation

Sivagowri Shanmugaratnam and Shivatharsiny Rasalingam

Abstract

Demand for environmental protection is gaining more public attention and legislative support. The development in industrial and technological sectors results in severe environmental issues, such as environmental contamination and energy shortage. Therefore, the development of new nanocomposites that can effectively act toward environmental remediation is necessary to overcome the detrimental environmental impacts. Transition metal chalcogenides (TMC) have gained worldwide attention in recent decades and are being researched for use in different applications due to their indirect bandgaps, optoelectronic behavior, and their stability that can enable the catalysts to absorb visible light that is abundant in solar radiation. In this chapter, synthesis, characterization, and application of TMCs, such as MS_x and MSe_x, toward environmental remediation application are reviewed. Efficiency of different TMC materials and different experimental conditions is also elaborated.

Keywords: transition metal chalcogenides, degradation, hydrothermal, organic compounds

1. Introduction

Effluents of industrial wastes are the important source of water pollution that are toxic in the environment and are becoming the biggest problem that influence the survival of human beings and other creatures in the environment. The demand of clean water is increasing with the rapid growth of global population; in the present, several treatment techniques that include biodegradation, membrane process, coagulation, adsorption, precipitation, sonochemical degradation, micellar-enhanced ultrafiltration, and advanced oxidation process (AOP) have been utilized to remove pollutants from wastewater [1]. Among them, AOP is an efficient method to remove the contaminants that are not degradable by means of biological processes. Since these processes involve the production of very reactive oxygen species (ROS), it can be able to destroy a wide range of organic compounds. A wide range of semiconductor photocatalysts have been utilized in AOPs. The first photocatalyst, TiO_2, discovered by Fujima and Honda in 1972 was found as gold standard due to its significant characteristics that include high chemical stability, nontoxicity, and relatively low price [2]. However, it only absorbs ultraviolet irradiation from sunlight due to its wide bandgap. To overcome this problem, several studies are focusing on finding ways to extend the absorption wavelength range of TiO_2. In this regard, metal or nonmetal elements doped onto TiO_2 and visible light response photocatalysts are being developed as the future generation. Among which transition metal chalcogenide (TMC) materials gained worldwide attention in recent decades because of their significant characteristics, such as excellent optical absorption due to its tunable indirect bandgap energies (1–2 eV) [3]. Moreover TMCs with graphene oxides [4], carbon nitride [5], metal oxides [6], and metals [7] were found to increase the conductivity of electrons, provide active sites, and effectively separate the electron and hole pairs generated by the semiconductor photocatalysts [4]. This chapter explores the synthesis, characterization, and the applications of TMCs.

2. Transition metal chalcogenides (TMCs)

Transition metal chalcogenides are considered as emerging candidates due to their unique physical and chemical properties and are being researched for use in lithium-ion batteries, solar cells, hydrogen evolution, and photocatalytic degradation due to their indirect bandgaps, optoelectronic behavior, and stability. In addition, nanodots (quantum dots)/nanostructures of these metal chalcogenides show stronger edge effects, and the quantum confinement effect makes it possible to utilize under solar-simulated irradiation [8, 9].

2.1 Different preparation methods

There are several methods to synthesize transitional metal chalcogenides, including ultrasonic chemical method, hydrothermal method, simple template-free one-pot method, an ion-exchange and precipitation methods, simple microwave-assisted solvothermal process, and surface modification method [5–12]. The following sections explore the most common preparation methods of TMCs.

Hydrothermal reaction usually occurs in the reaction vessel of stainless steel autoclave. The presence of aqueous solution or mineralizers under high pressure and low temperature encourages interaction of precursor materials during the process. This method has the feature to make high crystalline nanostructures or products at relatively low temperature [13, 14]. A range of TMCs that include NiS [15], CuS, FeS_2, NiS_2 [10], etc. have been synthesized using hydrothermal treatment method due to low temperature requirement (150°C), short time duration (~4 h), and high yield (>90%). In addition, there is no need to use any complexing agents or no need to control the pH [10] (**Figure 1**).

Further, hydrothermal method has been utilized to synthesize TMC-doped metal [16], metal oxides [17], or carbonaceous [18] materials. The following schematic illustration represents the preparation of CdS nanosheet-RGO hybrid material using hydrothermal treatment method (**Figure 2**).

Apart from these, microwave-assisted synthesis method has also been used to prepare TMCs with controllable size and shape compared with conventional heating method due to the homogeneous heating process. Further, this method can promote nucleation and, thus, reduce the time required for synthesis [18].

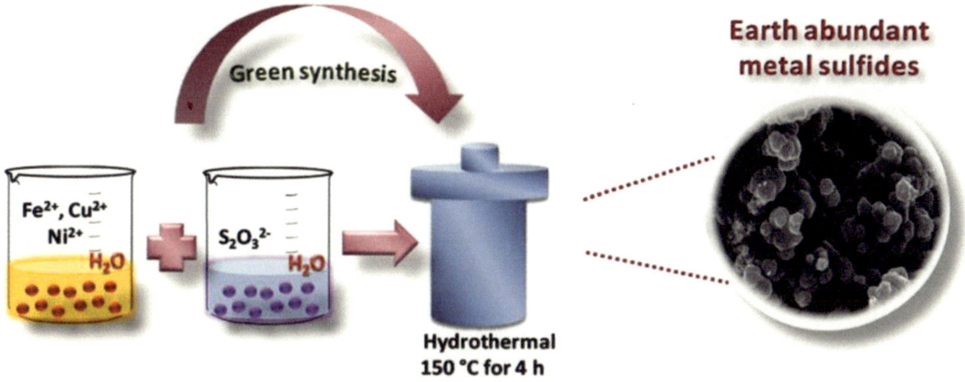

Figure 1.
Schematic illustration of the hydrothermal synthesis of FeS_2, CuS, and NiS_2 powders. Reproduced with permission from [10].

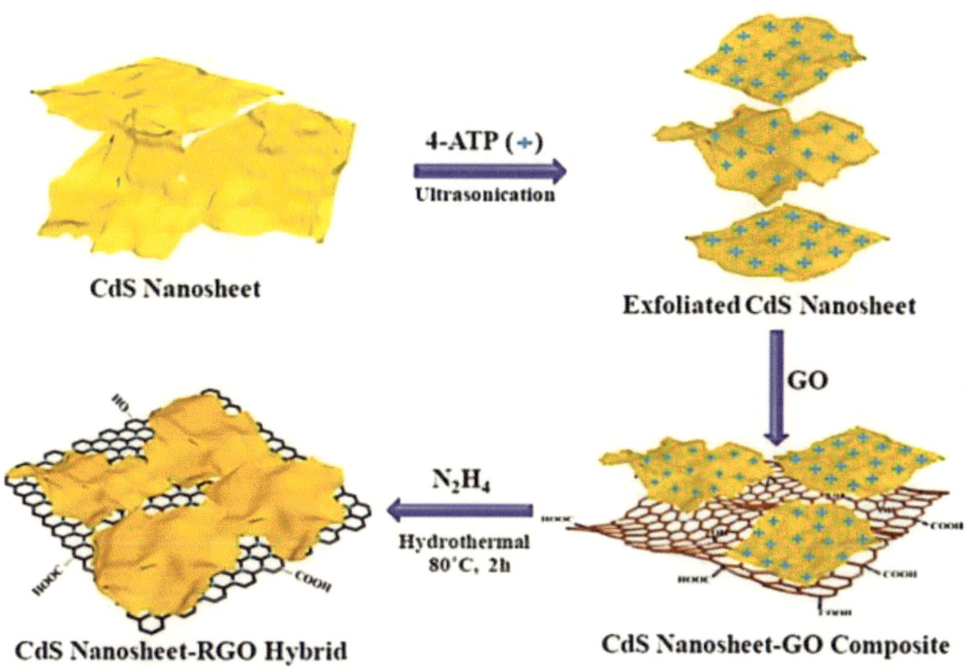

Figure 2.
Synthesis of CdS nanosheet-RGO hybrid. Reproduced with permission from [18].

Simple template-free one-pot method is a simple synthetic approach that was used to develop certain metal oxide hollow spheres based on direct solid evacuation arising from Ostwald ripening, the Kirkendall effect, or chemically induced self-transformation [19].

Figure 3 illustrates the mechanism for the simple template-free one-pot method to prepare hollow sphere, Ni-doped CdS material. Herein, biomolecular glutathione was used as sulfur and bubble source [7].

2.2 Characterization techniques for transition metal dichalcogenides

The essential techniques used to characterize the transition metal chalcogenide materials are discussed in this section.

2.2.1 X-ray diffraction spectroscopy

X-ray diffraction (XRD) spectroscopy is a nondestructive method used to analyze the crystallinity, crystal structure, crystallite size, and phase composition of the photocatalytic materials which presents in powder form or thin film.

In the study by Li and co-workers, MoS_2 nanosheets coupled into the carbon nitride to form MoS_2/C_3N_4 heterostructures synthesized by a facile ultrasonic chemical method were characterized by using XRD. All the samples prepared

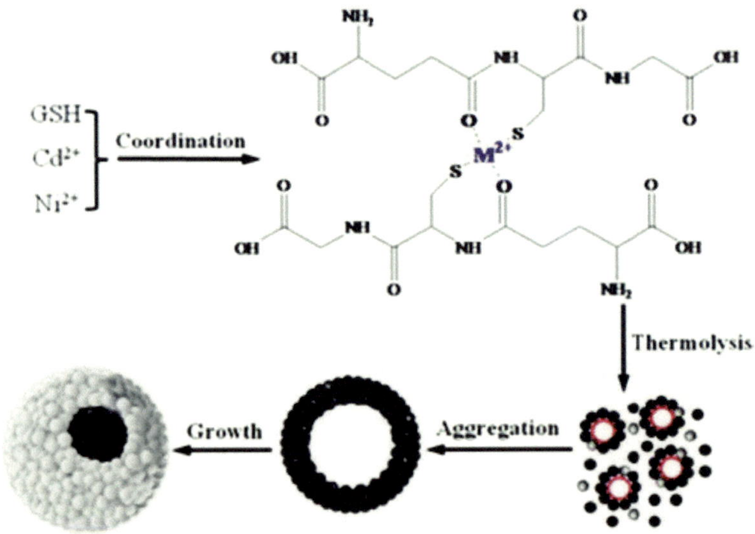

Figure 3.
Mechanism for the formation of Ni-doped CdS. Reproduced with permission from [7].

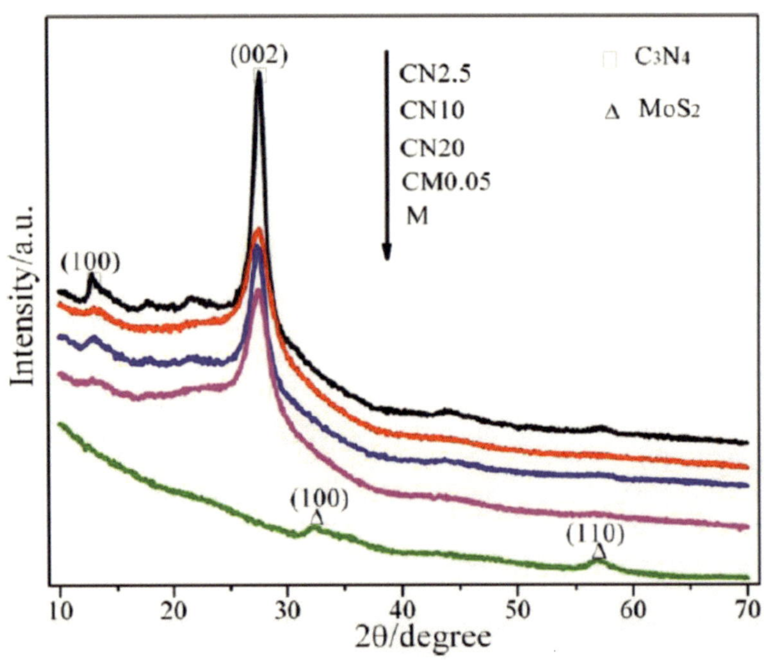

Figure 4.
XRD pattern for MoS_2 on g-C_3N_4 in different amounts. Reproduced with permission from [5].

showed similar patterns, and the peaks at 32.6 and 58.3° were assigned to the (100) and (110) crystal plane formation of MoS_2. Peak at 27.4° indicated to the (002) stacking layered structure, while the 13.0° peak corresponds to the (100) in-plane repeated units. The observation from XRD confirms the formation of MoS_2/C_3N_4 heterostrucuture [5] (**Figure 4**).

In another study, NiS nanoparticles prepared hydrothermally were characterized by powder XRD method.

Figure 5 illustrates the XRD patterns of as-prepared nanostructured NiS samples. The patterns were indexed to rhombohedral structured NiS with the space group of R3m. The cell parameters were found to be a = 9.61 Å and c = 3.16 Å [15].

XRD pattern (**Figure 6**) confirmed the formation of CdS, and Ni-doped CdS materials prepared by one-pot synthesis method show similar pattern and are assigned to hexagonal phase. It was also found that the peak intensity increases with the increase in Ni^{2+} doping concentration. Average crystallite size was estimated based on the broadening of the (002) peak, and it was found that the crystallite size also increased with the %Ni doped on CdS [7].

X-ray powder diffraction patterns of different metal sulfides were reported by Ali and co-workers as illustrated in **Figure 7**. The diffraction patterns of the samples are indexed to pyrite FeS_2, covellite CuS, and vaesite NiS_2. The purity of the phases was confirmed by the absence of additional peaks in their XRD. XRD patterns of cubic FeS_2 and NiS_2 were found to be isostructural, whereas CuS was found in a hexagonal symmetry. The crystallite sizes of the materials were estimated as 50, 45, and 22 nm for FeS_2, CuS, and NiS_2, respectively [10].

2.2.2 Electron microscopy

Scanning electron microscopy (SEM) is one of the electron microscopic techniques that images the sample surface using high beam electrons. In this technique, beam electrons strike the surface of the specimen and interact with atoms and form the secondary electrons, which characterized the information about the

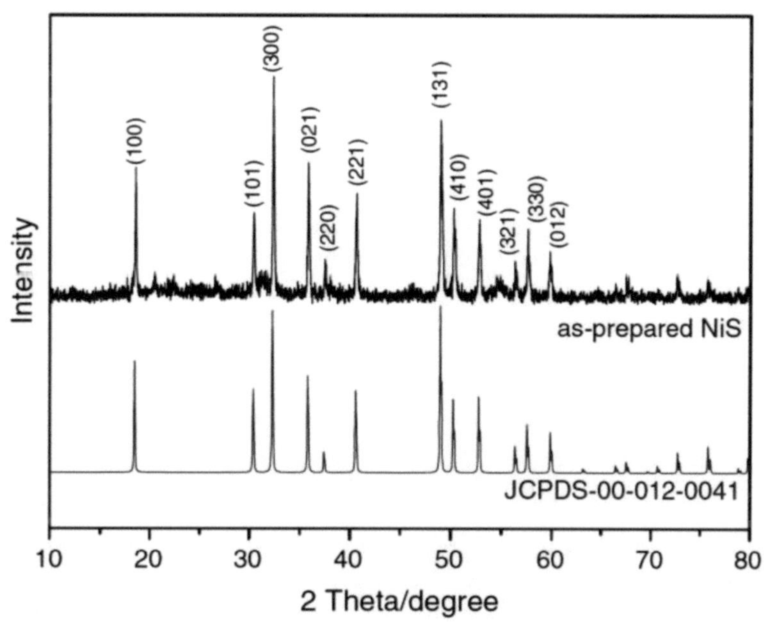

Figure 5.
XRD for synthesized NiS. Reproduced with permission from [15].

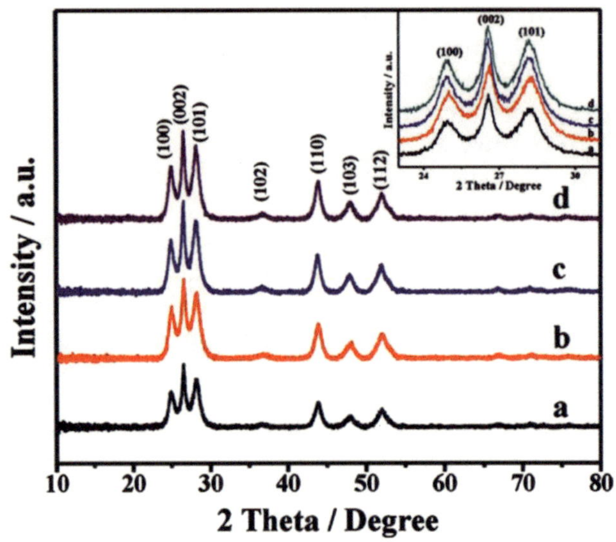

Figure 6.
XRD for Ni doped on CdS. Reproduced with permission from [7].

morphology of sample surface in high-resolution images colored in black and white. SEM is used to view dispersion of nanoparticles in different nanostructures, such as nanotubes, nanoclays, nanofillers, and nanofibers. In addition, the method used X-ray to identify elemental composition known as energy-dispersive ray method (EDX). The transmission electron microscopy (TEM) images form when beam of electron is transmitted through the sample and magnified by objective lens. This technique is the most preferred to measure the particle size and thickness of sample. In particular, for the morphology of materials, monitoring morphology and dispersion is very crucial compared to scanning electron microscopy [20].

Figure 8(a) shows SEM images of carnation flowerlike morphology with a diameter of about 500–700 nm. **Figure 8(b) and (c)** is the magnified view of an individual SnS_2 flower containing ultrathin nanopetals as building blocks, which is in 10 nm thick. These building blocks linked together and form 3D flowerlike structure. **Figure 8(d)** and **(e)** shows the TEM images of dense core of SnS_2 flower structure that clearly indicates that the nanopetals are well arranged across the flower center; however, the nanopetals of SnS_2 were not connected together. Finally, **Figure 8(f)** shows nanoparticles in randomly arranged manner with uneven sizes [21].

In the green synthesis of earth-abundant metal sulfides by Ali and co-workers, the samples prepared exhibited uniform morphology and particle size distribution, which is due to a controlled growth during the hydrothermal synthesis. The presence of iron, copper, nickel, sulfur, and carbon and its purity obtained using SEM analysis was found to be in good agreement with the phases observed by XRD. In addition, the atomic content of Fe, Cu, Ni, and S in the samples had a good correlation with theoretical atomic% of all the phases. The morphological analysis confirmed that as-prepared sulfides are in the form of agglomerated particles. In addition, polyhedral particles with the size of ~676 ± 44 nm were obtained for FeS_2 sample, whereas CuS and NiS_2 exhibited irregular- and spherical-shaped particles with the size of ~783 ± 53 and 933 ± 68 nm, respectively [10] (**Figure 9**).

Li et al. studied the morphology and microstructure of the as-prepared samples of WS_2 by using SEM and TEM. The detailed structural information of the WS_2/Bi_2MoO_6 composite was obtained by TEM analysis, and an irregular platelet-shaped nanostructure was observed with the length ranging from 500 nm to 2 μm.

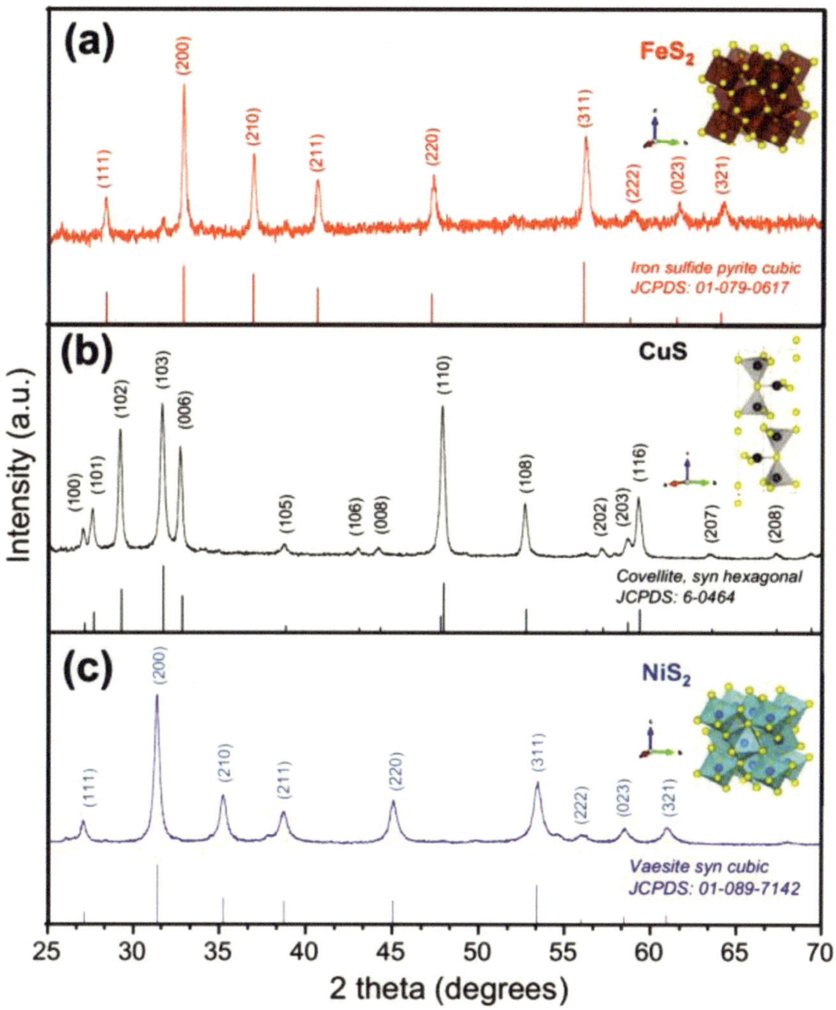

Figure 7.
XRD diffraction patterns and schematic representations of the crystalline structure of (a) FeS_2, (b) CuS, and (c) NiS_2. Reproduced with permission from [10].

The results obtained from TEM were in good agreement with SEM analysis. The HRTEM of the prepared materials further confirm the nanojunction structure. **Figure 10** illustrates two sets of different lattice images with the lattice fringe of 0.27 and 0.46 nm obtained from TEM corresponded to the (100) plane of WS_2 and (120) plane of Bi_2MoO_6 resulted from the XRD [17].

2.2.3 X-ray photoelectron spectroscopy

XPS or X-ray photoelectron spectroscopy is used as surface analytic technique based on the photoelectric effect. XPS is mainly used to find the composition of elements in the surface of the materials and to determine the valence band structure and chemical state of components.

Figure 11(a) illustrates the XPS molybdenum 3d spectra for $MoSe_2$, which exhibit peaks on 228.84 and 232.01 eV indicating binding energy of Mo $3d_{5/2}$ and Mo $3d_{3/2}$ for Mo^{4+} ions, respectively. In addition, **Figure 11(c)** depicts the XPS of $MoSe_X$ (at 120°C) indicating additional peaks at 232.07 and 229.45 eV which can be

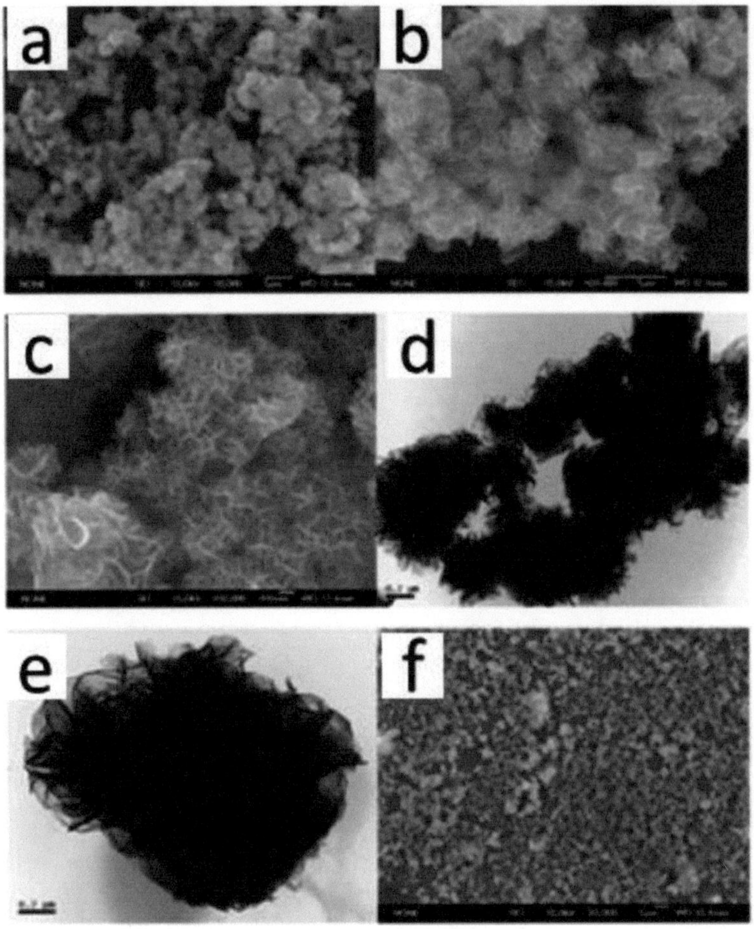

Figure 8.
(a–c) SEM images of SnS_2 flowerlike structure, (d–e) TEM images of SnS_2 flowerlike structure, and (f) SEM images of SnS_2 nanoparticles. Reproduced with permission from [21].

attributed to the binding energy of Mo^{5+} ions. The results confirmed the synthesis at 120°C, which resulted in two chemical states (IV and V) of molybdenum. Further, it was found that the amorphous $MoSe_2$ showed a very broad Se 3d peak than the crystalline $MoSe_2$. The Se $3d_{5/2}$ and Se $3d_{3/2}$ peaks for amorphous and crystal samples of $MoSe_2$ were found to lie between the range of 52 and 57 eV; however, the intensity of the peaks varied in amorphous sample. Peak-fitting analysis showed that the ratio of Mo to Se in the amorphous $MoSe_x$ sample was 1:3:1 [3].

Figure 12(a) shows the schematic of $PtSe_2$ in monolayer format, in that one sublayer of Pt atoms is sandwiched between two Se sublayers. **Figure 12(b)** shows the XPS measurements for Se during the growth of $PtSe_2$ at 270°C. The peaks at 55.68 and 54.80 eV indicate the binding energy of Se(0) chemical state at 25°C, and at 270°C, the peaks were found at 55.19 and 54.39 eV corresponding to the binding energy of Se^{2-} chemical state; in between that, at 200°C, the curve indicates the coexistence of Se(0) and Se^{2-} states. The binding energies of Se confirmed the formation of $PtSe_2$ at 270°C [22].

The XPS survey spectrum of SnS_2-AP (**Figure 13**) confirmed the presence of Sn and S components, whereas Cr, C, and O were found to be the contaminants. High-resolution XPS spectra of Sn 3d and S 2p core levels proved that the binding energies of Sn 3d and S 2p of SnS_2-AP were very close to those of SnS_2-(c). The binding

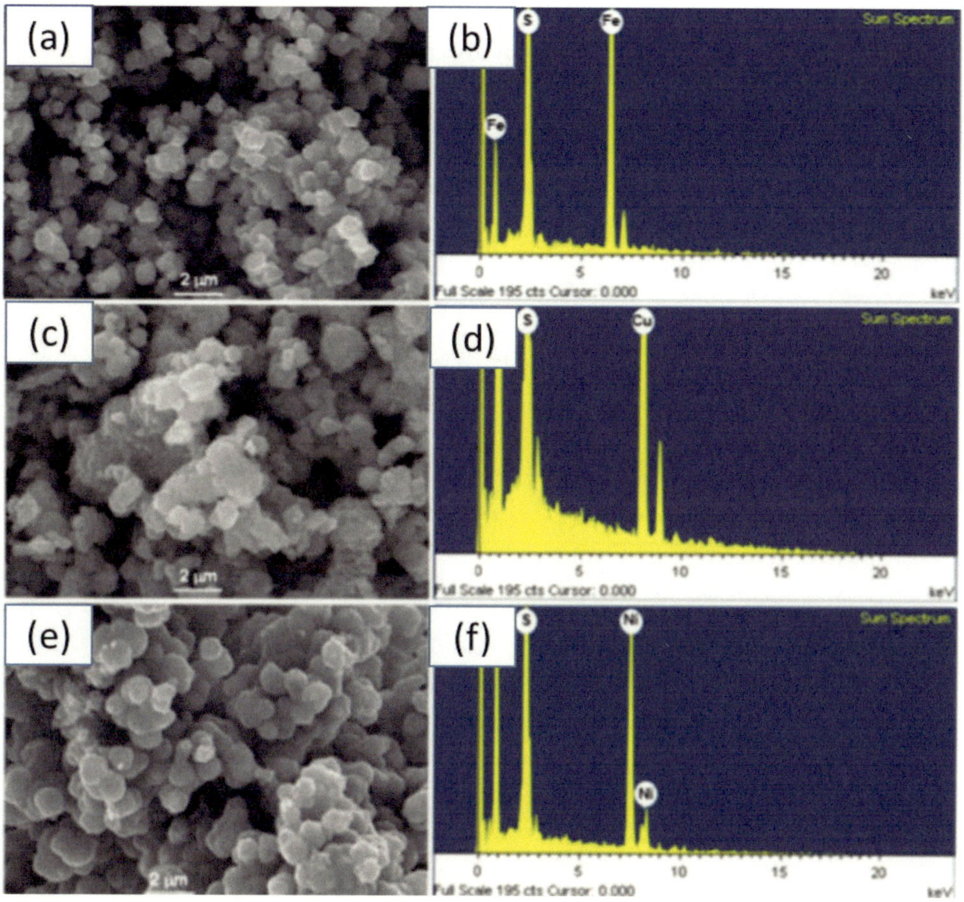

Figure 9.
SEM and EDS analysis of (a, b) FeS_2, (c, d) CuS, and (e, f) NiS_2. Reproduced with permission from [10].

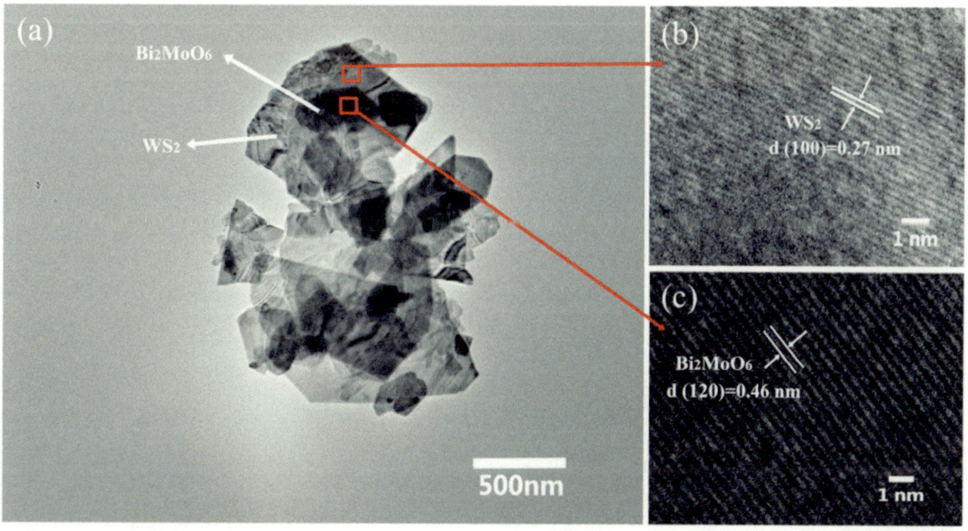

Figure 10.
TEM and HRTEM images of the hierarchical WS_2/Bi_2MoO_6 composite (5 wt% of WS_2). Reproduced with permission from [17].

energies of Sn $3d_{5/2}$ and S $2p_{3/2}$ of SnS_2-AP and SnS_2-(c) were 486.61 and 486.65, 161.68, and 161.74 eV, respectively. It was also found that the binding energies of Sn $3d_{5/2}$ and S $2p_{3/2}$ of SnS_2-AP and SnS2-(c) were of close proximity with the reference data of Sn^{4+} and S^{2-} in SnS_2.

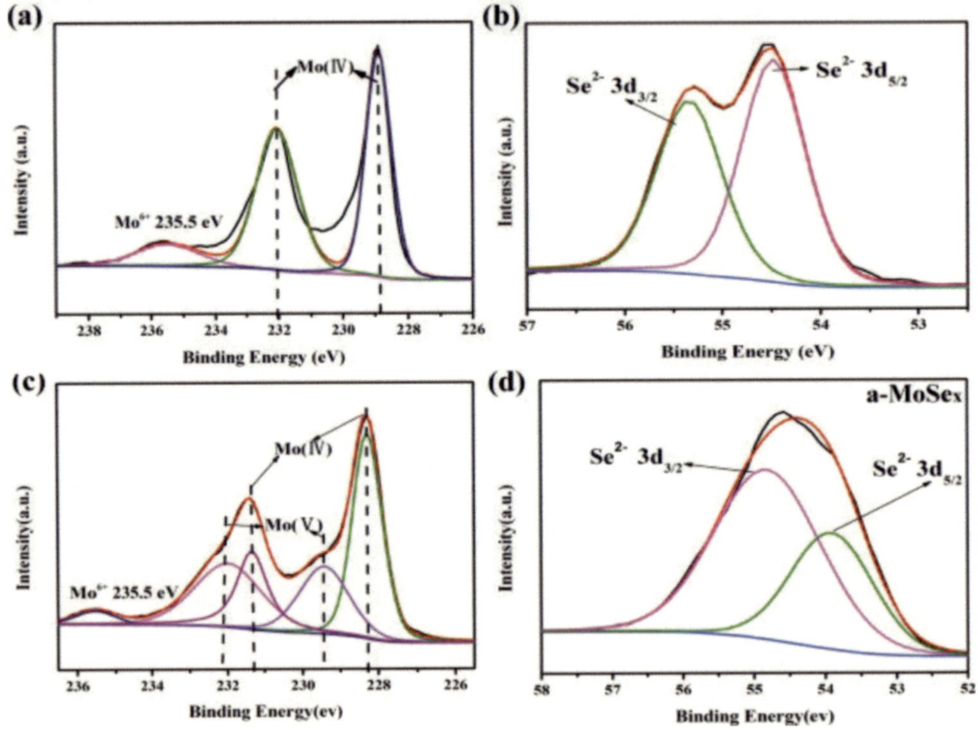

Figure 11.
(a, b) XPS for $MoSe_2$, (c, d) XPS for $MoSe_X$. Reproduced with permission from [3].

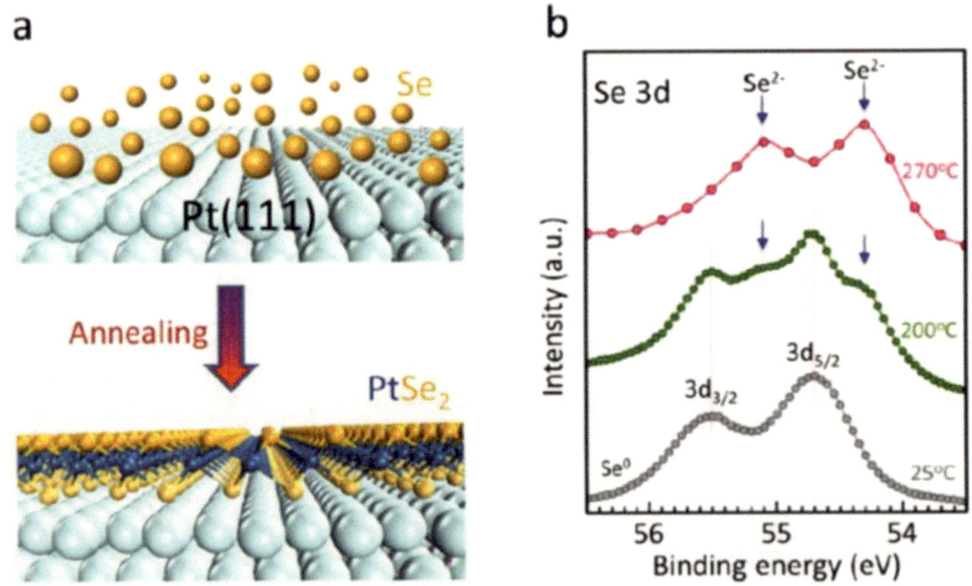

Figure 12.
(a) $PtSe_2$ monolayer and (b) XPS for $PtSe_2$. Reproduced with permission from [22].

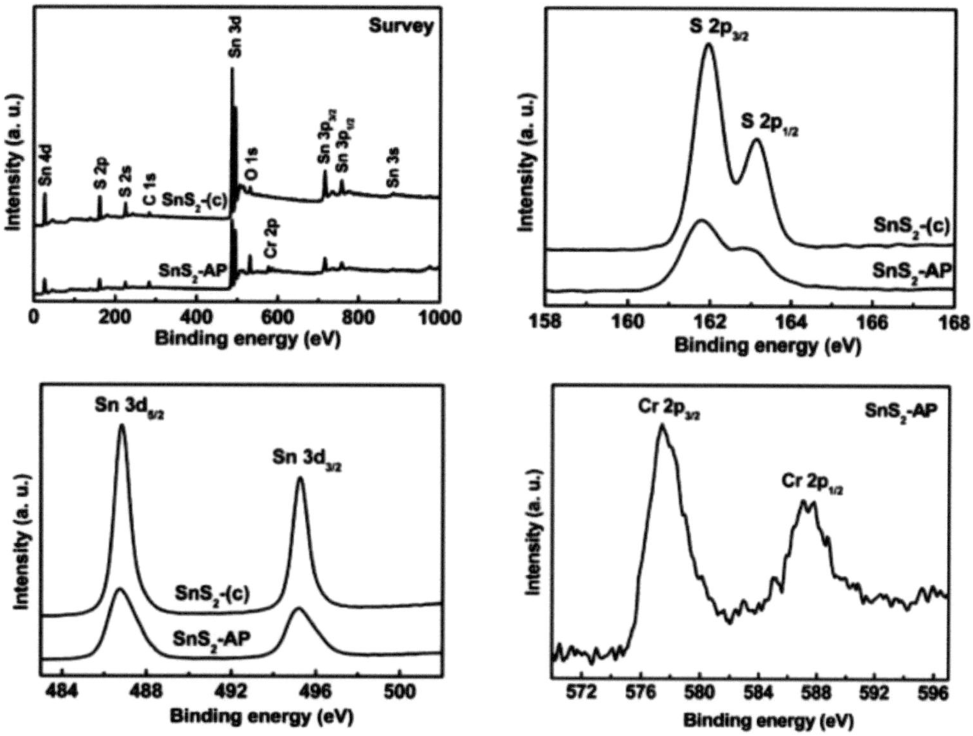

Figure 13.
XPS spectra of SnS_2-AP and SnS_2-(c). Reproduced with permission from [23].

Apart from these, a peak attained at 577.36 eV is due to the binding energy of Cr $2p_{3/2}$ corresponding to Cr (III) in $Cr(OH)_3$. The hydrolysis-precipitation of Cr(III) ion leads to the formation of $Cr(OH)_3$ on the surface of SnS_2-AP. Cr(III) were generated from the photocatalytic reduction of adsorbed Cr(VI) [23].

2.3 Photodegradation of organics using TMCs

Persistent organic pollutants (POPs) are of global concern because of their potential for long-range transport, persistence in the environment, ability to biomagnify and bioaccumulate in ecosystems, as well as possess negative effects on human health and the environment. In particular, the agrochemicals and textile effluents involve processes that produce compounds that are very toxic to the environment. Thus, it is necessary to remove these colored pollutants from the environment. Several catalysts have been utilized to remove the pollutants from the polluted environment, and the transition metal chalcogenide has recently gained much attention in this regard due to their significant characteristic properties.

Wen-chao Peng et al. synthesized MoS_2/reduced graphene oxide hybrid with CdS nanoparticle for photocatalytic reduction of nitroaromatic compounds to aromatic amines under visible light irradiation in the presence of sacrificial agent. It was also noted that composite of CdS-0.03 (MoS_2/0.01rGO) exhibited remarkable enhancement on the 4-nitrophenol reduction due to the separation of electron-hole pairs generated by CdS nanoparticles [4] (**Figure 14**). In addition, it was found that the photocatalyst appeared to be stable confirmed by recycling study as illustrated in the following figure.

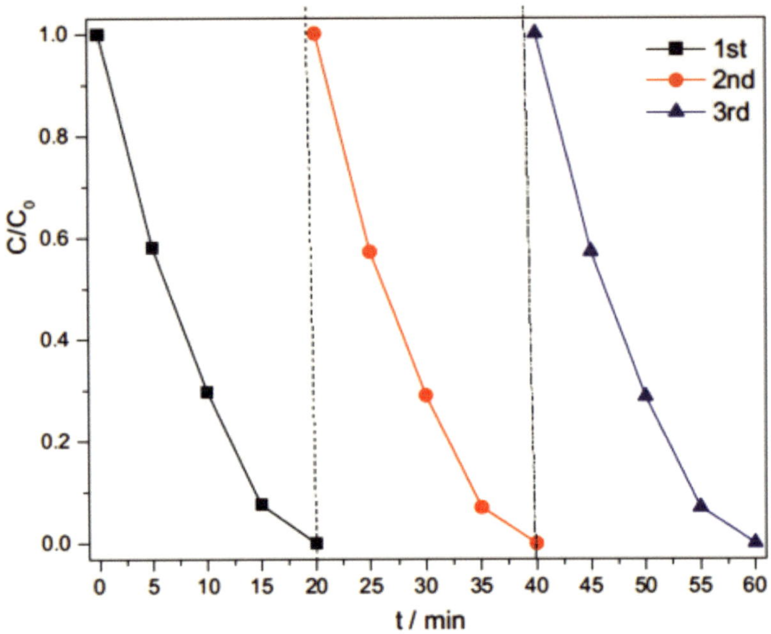

Figure 14.
Repeated tests using recycled CdS 0.03(MoS₂/0.01rGO) composite catalyst for photocatalytic 4-nitrophenol reduction. Reproduced with permission from [4].

In another study, Peitao Liu and co-workers worked on prepared N-doped MoS_2 nanoflowers (specific surface area of 114.2 $m^2\ g^{-1}$) and α-Fe_2O_3@N-doped MoS_2 heterostructures for decoloring or the removal of the Rhodamine B (RhB) under visible light irradiation. When comparing the activity of these two materials, N-doped MoS_2 nanoflowers showed optimum rate constant as 0.0928 min^{-1}, which was 26.4 times greater than that of MoS_2 nanosheets. Further, 20 mg of these catalysts completely degraded the 50 mL of 30 mgL^{-1} RhB within 70 min [24] (**Figure 15**).

MoS_2 nanosheets were coated into carbon nitride synthesized by facile ultrasonic chemical method to form heterostructures. These materials were used to explore the photocatalytic degradation of the dyes, RhB and MO, and the optimum reaction rate constant was found to be 301 min^{-1} with the 0.05 wt% MoS_2 [5].

Yangyang Liu and co-workers synthesized ZnO nanosheet doped with P by using conventional chemical vapor transport and condensation (CVTC) method and then coated with an atomic layer of MoS_2 on it (**Figure 16**).

The photocatalytic activity of the catalyst was tested by degradation of organic dyes, such as methylene blue (MB) and Rhodamine B (RhB), under natural sunlight. It was found that 95% of organic dyes were degraded within 6 min and photocatalytic rate constant was found to be 1.413 min^{-1}, which was 3.4 times better than that of P25 due to the enhanced light adsorption efficiency obtained by synergetic effect of MoS_2 [6].

In another study, Haiyang Liu et al. worked on 1 T@2H-MoS_2/Ag composite synthesized by microwave-hydrothermal and photoreduction methods for photocatalytic degradation of conventional dyes and photocatalytic reduction of Cr(VI) under visible light irradiation. It was observed that 1 T@2H-MoS_2/Ag effectively enhanced photocatalytic activity compared with 2H-MoS_2 due to the enhancement in the light response range and charge separation by using Ag quantum dots and 1 T phase. In addition, 1 T@2H-MoS_2 showed 81 and 41% of increment in

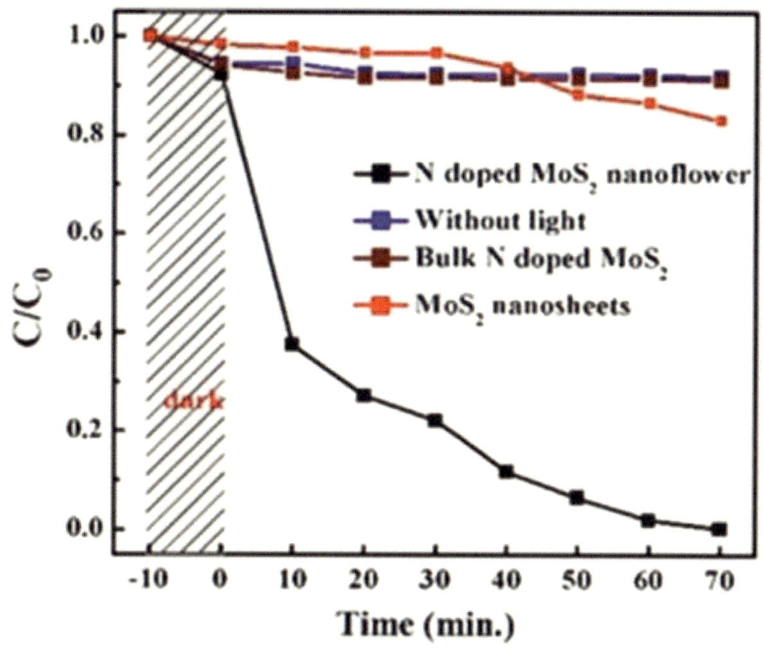

Figure 15.
Photocatalytic degradation of RhB by different photocatalysts (N-doped MoS_2 nanoflower, without light, bulk N-doped MoS_2, and MoS_2 nanosheets) under visible light irradiation, where C and C_0 are the RhB concentrations at time t min. And 0 min, respectively. Reproduced with permission from [24].

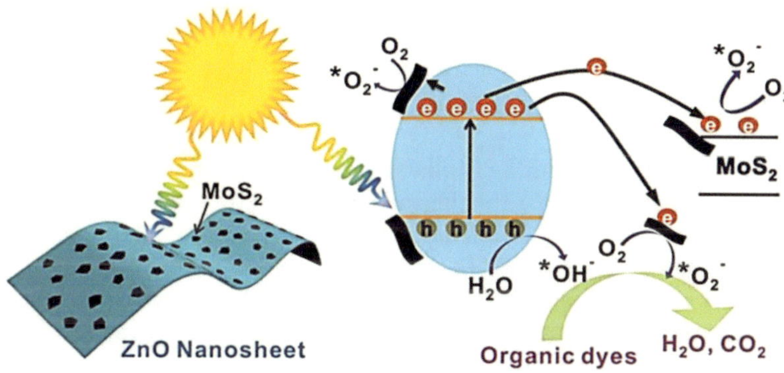

Figure 16.
Mechanism of photolysis on ZnO decorated with MoS_2. Reproduced with permission from [6].

photocatalytic reduction of Cr(VI) and photocatalytic degradation of methylene blue, respectively [16] (**Figure 17**).

In a different study, Xi Yang et al. worked on amorphous and crystalline molybdenum selenide synthesized by facile low-temperature hydrothermal method for adsorption and degradation of Rhodamine and methylene blue under dark and visible light irradiation. Due to the excellent optical absorption and narrow bandgap (1–2 eV), molybdenum selenide ($MoSe_2$) is preferred for the photocatalytic degradation of pollutants. Hole and free radical studies concluded that the photocatalytic effect of the amorphous samples was higher than that of the crystalline samples due to the different active substances. The active substances, superoxide radicals, in particular, predominated the process of degrading the dye. When comparing the amorphous and crystalline materials, amorphous molybdenum selenide showed

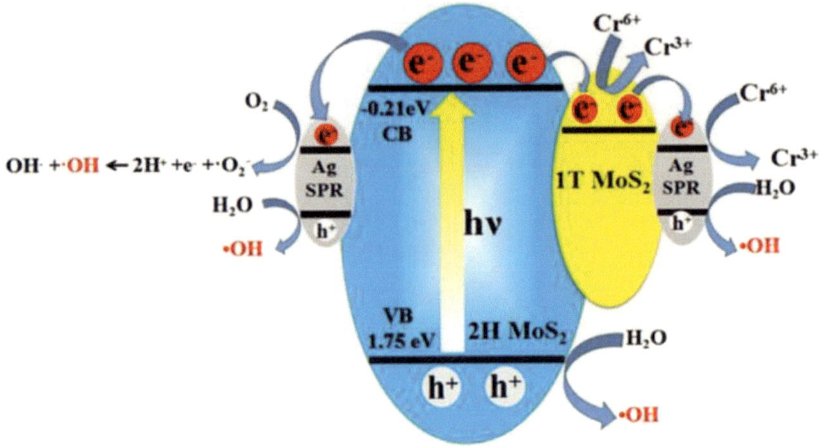

Figure 17.
Schematic illustration of photo-charge separation through 1 T@2H-MoS$_2$/Ag under visible light irradiation. Reproduced with permission from [16].

better performance in adsorption and degradation process due to the formation of superoxide radical ($^{\cdot}O^{2-}$) and more unsaturated atoms and greater specific surface area than crystalline structure; further, the holes and hydroxyl radicals were found to be the main active substances [3].

Hongxu Guo et al. synthesized NiS nanoparticles for photocatalytic degradation and adsorption of Congo red under visible light irradiation. NiS nanoparticles were synthesized by hydrothermal method, and it exhibited efficient photocatalytic activity. 30 mg/L of Congo red was completely degraded after illumination of 210 min. In addition, OH radicals were identified by using fluorescence technique, and those highly reactive radicals were found to be the major factor for the photocatalytic process [15].

In a different study, Aniruddha Molla et al. synthesized Ag-In-Ni-S nanocomposites with different shapes for the degradation of methylene blue under dark and visible light irradiation. It is noteworthy to mention here that 20 mg of the catalyst showed a complete degradation of methylene blue within 12 min, when there was no light exposure. Notably, the degradation took only 2 min, when visible light was exposed with a 100 W lamp source. Further, the reactive oxygen species generated during the process was confirmed in their study [2].

CdS and Ni-doped CdS hollow spheres were used by Man Luo and his group for the degradation or removal of RhB and phenol. These materials were synthesized by simple template-free one-pot method. 1.2 mol% of Ni-doped CdS hollow spheres exhibited better performance in the removal of organic pollutants due to the reduce recombination rate of the electrons and hole pairs [7]. **Figure 18** illustrates the UV-absorbance profile for the degradation of RhB dye with the 1.2 mol% of Ni-doped CdS.

Earth-abundant transition metal sulfides, such as FeS_2, CuS, and NiS_2, were synthesized by fast and low-cost hydrothermal synthesis method and utilized for photocatalytic hydrogen evolution and photocatalytic degradation of indigo carmine dye under visible light irradiation. Different sacrificial agents, such as Na_2S/Na_2SO_3, EDTA, and ethanol, were used when the H_2 evolution was tested. FeS_2 showed higher activity for the hydrogen evolution (32 $\mu molg^{-1} h^{-1}$), and an 88% of dye (indigo carmine) degradation was attained due to their suitable electronic and optical properties. The better activity was credited to a larger crystallite size, smaller particle size, and lower recombination rate of FeS_2 than other materials

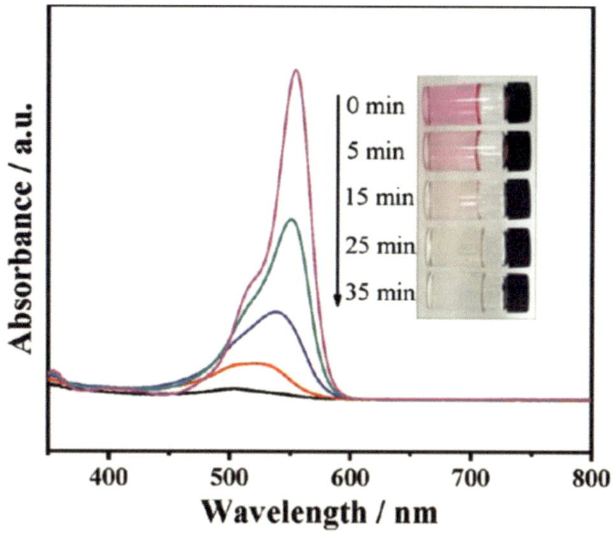

Figure 18.
UV-visible absorption spectra of RhB in the presence of 1.2 Mol% of Ni-doped CdS. Reproduced with permission from [7].

that include CuS and NiS_2. In addition, the presence of Na_2S/Na_2SO_3 exhibited higher enhancement in the activity for the production of hydrogen among the other scavenging reagents used in this study [10].

In another study, Yongje Zhao and co-workers utilized the bimetallic chalcogenides material, $FeCoS_2$, synthesized by using a modulated hydrothermal method for photodegradation of methylene blue and anode materials for Na-ion batteries under UV radiation. The photocatalytic activity of $FeCoS_2$ which was hydrothermally treated at 190°C was found to be a better candidate with the degradation efficiency of 90% (in 150 min) under UV irradiation. The better photocatalytic activity was related with the microstructure, specific surface area, and charge separation of catalytic materials [25].

Alireza et al. prepared NiS-clinoptilolite zeolite as a catalyst via an ion-exchange and precipitation methods. The catalyst was utilized for photodegradation of furfural in aqueous solution over UV irradiation. An optimum photocatalytic efficiency was observed when 330 mg L^{-1} of the catalyst was used at pH 5 in 6 mM of furfural solution. In addition, it was found that the degradation rate increased with increasing amount of hydrogen peroxide and potassium bromate in the solution [12].

Xi Li et al. used SnS_2 and CdS for photocatalytic degradation of different types of organic dyes, such as methyl orange (MO), Rhodamine B (RhB), Congo red (CR), orange II (OII), malachite green (MG), and methylene blue (MB), under visible light irradiation, and the activity of these two materials were compared. SnS_2 showed higher activity in the degradation of MO and lower activity in the degradation of RhB-like organic dyes, when compared to CdS material. In the degradation of azo dyes (dyes containing N=N double bond), SnS_2 followed a reduction mechanism with photoelectrons via the SnIV/SnII transition, whereas an oxidation mechanism was observed with both SnS_2 and CdS for photodegradation of organic dyes which do not consist of N=N double bond. $^{\cdot}O_2^{-}$ and $^{\cdot}OH$ radicals were found to be the major contributors for the oxidation mechanism. Since the reduction of dye molecules was faster than migration of the radicals, SnS_2 exhibited much higher activity than the CdS for the dyes containing N=N double bond. However, SnS_2 showed lower efficiency than CdS in the photocatalytic degradation of other organic dyes without N=N double bond, due to the slow production of reactive oxidative species mentioned above. Additional experiments were carried out in the presence of the scavengers NO, N_2, O_2, and IPA [26]. **Figure 19** depicts the rate of degradation of both catalysts against the azo and non-azo dyes.

In a different study, 3D carnation flowerlike hexagonal SnS_2 nanostructures were synthesized by a simple microwave-assisted solvothermal process for photocatalytic degradation of Rhodamine (RhB) and phenol under visible light irradiation. The 3D carnation flowerlike hexagonal SnS_2 nanostructures exhibited enhanced photocatalytic

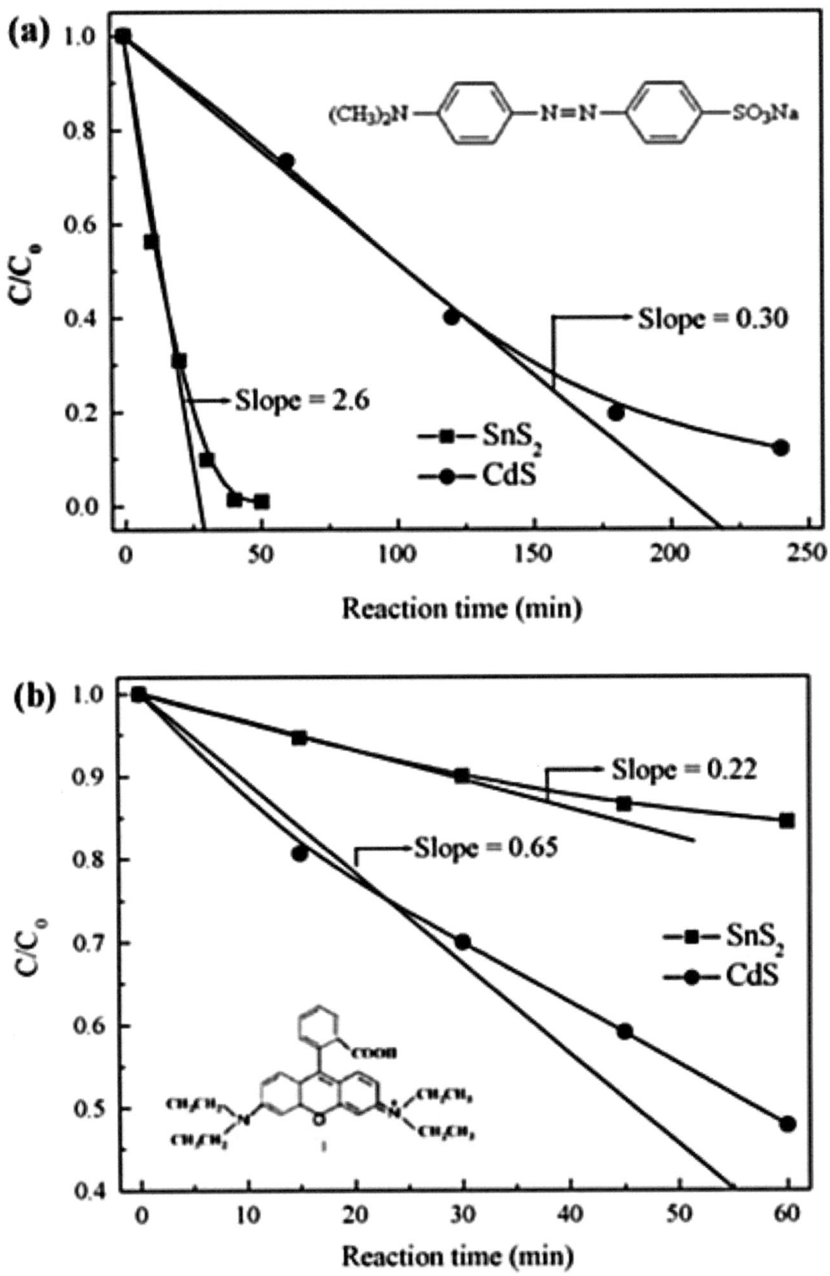

Figure 19.
Plots of normalized concentration (C_o/C) of MB vs. different time intervals (a) MO and (b) RhB photocatalytic degradation. Reaction conditions: 20 mg (a) or 50 mg (b) photocatalyst, 50 mL 6.0 × 10^{-5} Mol/L organic dyes, a 300 W xenon lamp(>420 nm), reaction temperature = 30°C. Reproduced with permission from [26].

activity compared to SnS_2 nanoparticles due to its high BET surface area, high surface-to-volume ratios, and increased light absorbance of hierarchical mesoporous structures [21].

Yong Cai Zhang et al. synthesized SnS_2 nanoflakes from $SnCl_2.H_2O$ for photocatalytic degradation of methyl orange (MO) under both the visible light and real sunlight irradiation. The SnS_2 material synthesized at 200°C exhibited high photocatalytic efficiency than others (i.e., 100% of degradation of MO after 60 min in the first cycle and 86% in the fifth cycle). The reason for the better degradation efficiency correlated with the combined action of several factors that include band-gap (2.21–2.25 eV), size, dispersibility, suspensibility, surface area, crystallinity, crystal defects, and photochemical stability. Almost all the materials showed better efficiency than the P25 material used for the comparison purpose [27].

SnS_2 nanocrystals with adjustable sizes were synthesized by Yong Cai Zhang and co-workers by utilizing hydrothermal method with the precursor of $SnCl_4.5H_2O$. The materials were tested for the reduction of aqueous Cr(VI) under visible light irradiation (**Figure 20**).

It was found that the reduction of aqueous Cr(VI) depends on their hydrothermal conditions. Particularly, SnS_2, synthesized at 150°C for 12 hours, showed highest photocatalytic activity in reducing aqueous Cr(VI) [10]. In a separate study, the same group worked on size-controlled synthesis of SnS_2 by using hydrothermal method, and their photocatalytic activity tested against the degradation of methyl orange under visible light irradiation [23].

Zhenyi Zhang et al. synthesized nanosheets of SnS_2 on g-C_3N_4 as 2D/2D heterojunction photocatalyst for the photocatalytic degradation of organics including dye (RhB) and phenols over visible light irradiation. It was found that the hetero-samples exhibited an enhancement in the photocatalytic degradation of organics than pure g-C_3N_4 and SnS_2 nanosheets. In this study, a better efficiency was attained for the catalyst loaded with 5.0 wt% SnS_2 (rate constant of ~0.2 min^{-1}) for the RhB photodegradation [28].

Atkin and co-workers utilized two-dimensional tungsten disulfide (WS_2) nanoflakes, hybridized with carbon dots prepared using two-step method for

Figure 20.
Photodegradation of Cr(VI) in the presence of SnS_2. Reproduced with permission from [23].

photocatalytic degradation of organic dyes. This material shows photocatalytic efficacy for the degradation of dyes. Twelve percent of degradation was obtained against the Congo red dye when 0.24 mg/L photocatalyst was utilized [29].

In another study, few layered WS_2/Bi_2MoO_6 heterojunction composites were prepared by Xiang Li et al. using hydrothermal method for photodegradation of organics under visible light irradiation. A higher degradation efficiency was attained by the materials further proved by the percentage removal of 99.5 and 91.7% for Rhodamine B (RhB), 98.9 and 89.8% for ciprofloxacin (CIP), 76.0 and 67.8% for methylene blue (MB), and 69.3 and 58.6% for methimazole (MMI), respectively. The reason for the removal efficiency was correlated with the structure of the materials, which may provide larger contact area for interfacial charge transfer and can shorten the migration distance of charge transfer [17] (**Figure 21**).

Hua-Bin Fang and co-workers used WS_2 nanosheets on manganese oxide (MnO_X) prepared by photo-deposition in $MnSO_4$ solution. Here, manganese oxide (MnO_X) was used as a hole-trapping material. MnO_X/WS_2 showed better performance in the degradation of RhB under visible light than pure WS_2, because of the efficient charge separation in the composite promoted by MnO_x [30].

Prabhakar Vattikuti et al. synthesized 1D Bi_2S_3 nanorod/2D e-WS_2 nanosheet heterojunction by hydrothermal method for the degradation of methyl orange (MO) dye in aqueous solution and evolution of hydrogen under visible light irradiation. The material showed enhanced photocatalytic activity due to the presence of strong surface active sites and fast transfer of electron–hole pairs in the heterostructure of the composite. In addition, 84.4% of methyl orange was degraded after 90 min of exposure of visible light irradiation when 3% of Bi_2S_3/e-WS_2 composite was used [31].

Rajesh Bera et al. synthesized CdS nanosheet (0D, 1D, and 2D) on graphene oxide composite by surface modification method. The catalysts prepared were utilized for the photocatalytic degradation of 4-aminothiophenol under visible light. The photocatalytic activity of CdS nanosheet/RGO composite was found to nearly 4 times, 3.4 times, and 2.5 times higher than CdS nanoparticle/RGO, CdS nanorod/RGO, and pure CdS nanosheet samples, respectively. In addition, 2D-2D nanoarchitecture was found to more effective than 0D-2D and 1D-2D hybrid systems due to its better ability for harvesting of photon from sunlight and transport of excitons to their reaction sites than others [18].

Sohrabnezhad et al. worked on CoS nanoparticles supported on Al-MCM-41 material, which was synthesized by ion-exchange method. The materials were

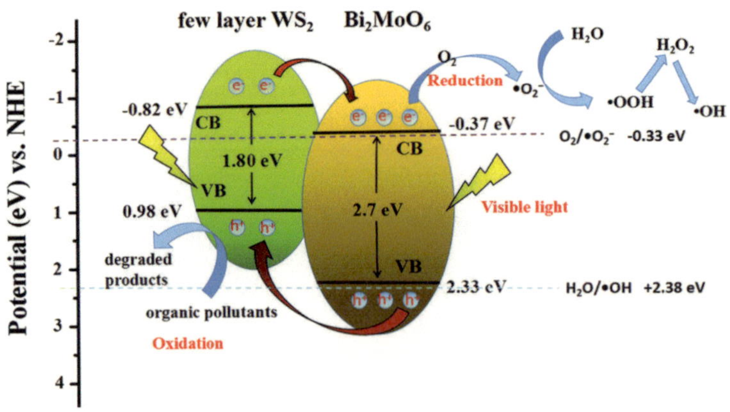

Figure 21.
Mechanism for the photocatalytic activity of WS_2/Bi_2MoO_6 in the presence of organic pollutants. Reproduced with permission from [17].

utilized for photocatalytic degradation of basic blue 9 or methylene blue (MB). Seven percent loading of CoS on the mesoporous support exhibited maximum efficiency for the photocatalytic degradation of methylene blue, and the optimum pH was found to be 9. The degradation followed pseudo-first-order reaction with the rate constant (k) of 0.0361. The effect of pH, amount of photocatalyst, and initial concentration of dye were also examined in their studies [11].

3. Conclusions

Metal sulfides/selenites (metal chalcogenides) have attracted considerable interest due to their electronic and optical applications. In addition, most of the common metal sulfides are nontoxic with narrow bandgap. Due to these characteristic features, metal chalcogenides have been utilized as potential candidates for photocatalysis. This chapter reviewed the most common synthetic protocol of several transition metal chalcogenides that include metal sulfides and metal selenites. Further, a brief section covered the basic characterization techniques, which have been commonly used to characterize the chalcogenide materials. Most importantly, the application of different metal chalcogenides toward environmental remediation applications, i.e., degradation of organic pollutants, is also explored herein.

Acknowledgements

The authors would like to thank the Department of Chemistry, University of Jaffna, Jaffna, Sri Lanka.

Conflict of interest

The authors declare no conflict of interests.

Author details

Sivagowri Shanmugaratnam and Shivatharsiny Rasalingam*
Department of Chemistry, University of Jaffna, Jaffna, Sri Lanka

*Address all correspondence to: srtharsha12@gmail.com

References

[1] Rasalingam S, Peng R, Koodali RT. Removal of hazardous pollutants from wastewaters: Applications of TiO_2-SiO_2 mixed oxide materials. Journal of Nanomaterials. 2014;**2014**:10. DOI: 10.1155/2014/617405

[2] Molla A, Sahu M, Hussain S. Under dark and visible light: Fast degradation of methylene blue in the presence of Ag–In–Ni–S nanocomposites. Journal of Materials Chemistry A. 2015;**3**(30): 15616-15625. DOI: 10.1039/c5ta02888d

[3] Yang X et al. Amorphous molybdenum selenide as highly efficient photocatalyst for the photodegradation of organic dyes under visible light. Applied Surface Science. 2018;**457**: 214-220. DOI: 10.1016/j.apsusc.2018.06.039

[4] Wc P, Chen Y, Li X. MoS_2/reduced graphene oxide hybrid with CdS nanoparticles as a visible light-driven photocatalyst for the reduction of 4-nitrophenol. Journal of Hazardous Materials. 2016;**309**:173-179. DOI: 10.1016/j.jhazmat.2016.02.021

[5] Li Q et al. High efficiency photocatalysis for pollutant degradation with MoS_2/C_3N_4 heterostructures. Langmuir. 2014;**30**(29):8965-8972. DOI: 10.1021/la502033t

[6] Liu Y et al. A highly efficient sunlight driven ZnO nanosheet photocatalyst: Synergetic effect of P-doping and MoS_2 atomic layer loading. ChemCatChem. 2014;**6**(9):2522-2526. DOI: 10.1002/cctc.201402191

[7] Luo M et al. One-pot synthesis of CdS and Ni-doped CdS hollow spheres with enhanced photocatalytic activity and durability. ACS Applied Materials & Interfaces. 2012;**4**(3):1813-1821. DOI: 10.1021/am3000903

[8] Zheng L, Zhang W, Xiao X. Preparation of titanium dioxide/tungsten disulfide composite photocatalysts with enhanced photocatalytic activity under visible light. Korean Journal of Chemical Engineering. 2016;**33**(1):107-113. DOI: 10.1007/s11814-015-0098-7

[9] Wang Q et al. MoS_2 quantum dots @ TiO_2 nanotube arrays: An extended-spectrum-driven photocatalyst for solar hydrogen evolution. ChemSusChem. 2018;**11**(10):1708-1721. DOI: 10.1002/cssc.201800379

[10] Huerta-Flores AM et al. Green synthesis of earth-abundant metal sulfides (FeS_2, CuS, and NiS_2) and their use as visible-light active photocatalysts for H_2 generation and dye removal. Journal of Materials Science: Materials in Electronics. 2018;**29**(13):11613-11626. DOI: 10.1007/s10854-018-9259-x

[11] Sohrabnezhad S, Pourahmad A, Radaee E. Photocatalytic degradation of basic blue 9 by CoS nanoparticles supported on AlMCM-41 material as a catalyst. Journal of Hazardous Materials. 2009;**170**(1):184-190. DOI: 10.1016/j.jhazmat.2009.04.108

[12] Nezamzadeh-Ejhieh A, Moeinirad S. Heterogeneous photocatalytic degradation of furfural using NiS-clinoptilolite zeolite. Desalination. 2011;**273**(2-3):248-257. DOI: 10.1016/j.desal.2010.12.031

[13] S-H Y. Hydrothermal/solvothermal processing of advanced ceramic materials. Journal of the Ceramic Society of Japan. 2001;**109**(1269): S65-S75. DOI: 10.2109/jcersj.109.1269_S65

[14] Kumar A, Pandey G. Different methods used for the synthesis of

TiO_2 based nanomaterials: A review. American Journal of Nano Research and Applications. 2018;**6**(1):1. DOI: 10.11648/j.nano.20180601.11

[15] Guo H et al. Efficient adsorption and photocatalytic degradation of Congo red onto hydrothermally synthesized NiS nanoparticles. Journal of Nanoparticle Research. 2013;**15**(3):1475. DOI: 10.1007/s11051-013-1475-y

[16] Liu H et al. Synergetic photocatalytic effect between 1 T@ 2H-MoS_2 and plasmon resonance induced by Ag quantum dots. Nanotechnology. 2018;**29**(28):285402. DOI: 10.1088/1361-6528/aabf56

[17] Li X et al. Fabrication of a novel few-layer WS_2/Bi_2MoO_6 plate-on-plate heterojunction structure with enhanced visible-light photocatalytic activity. Dalton Transactions. 2018;**47**(30):10046-10056. DOI: 10.1039/C8DT02109K

[18] Bera R, Kundu S, Patra A. 2D hybrid nanostructure of reduced graphene oxide–CdS nanosheet for enhanced photocatalysis. ACS Applied Materials & Interfaces. 2015;**7**(24):13251-13259. DOI: 10.1021/acsami.5b03800

[19] Cheng B et al. One-pot template-free hydrothermal synthesis of monoclinic hollow microspheres and their enhanced visible-light photocatalytic activity. International Journal of Photoenergy. 2012. DOI: 10.1155/2012/797968

[20] Joshi M et al. Characterization techniques for nanotechnology applications in textiles. Indian Journal of Fibre and Textile Research. 2008;**33**(3):304-317

[21] Liu H et al. Microwave-assisted solvothermal synthesis of 3D carnation-like SnS_2 nanostructures with high visible light photocatalytic activity. Journal of Molecular Catalysis A: Chemical. 2013;**378**:285-292. DOI: 10.1016/j.molcata.2013.06.021

[22] Wang Y et al. Monolayer $PtSe_2$, a new semiconducting transition-metal-dichalcogenide, epitaxially grown by direct selenization of Pt. Nano Letters. 2015;**15**(6):4013-4018. DOI: 10.1021/acs.nanolett.5b00964

[23] Zhang YC et al. Size-tunable hydrothermal synthesis of SnS_2 nanocrystals with high performance in visible light-driven photocatalytic reduction of aqueous Cr(VI). Environmental Science & Technology. 2011;**45**(21):9324-9331. DOI: 10.1021/es202012b

[24] Liu P et al. Flower-like N-doped MoS_2 for photocatalytic degradation of RhB by visible light irradiation. Nanotechnology. 2016;**27**(22):225403. DOI: 10.1088/0957-4484/27/22/225403

[25] Zhao Y et al. The synthesis of $FeCoS_2$ and an insight into its physicochemical performance. CrystEngComm. 2018;**20**(15):2175-2182. DOI: 10.1039/C8CE00299A

[26] Li X, Zhu J, Li H. Comparative study on the mechanism in photocatalytic degradation of different-type organic dyes on SnS_2 and CdS. Applied Catalysis B: Environmental. 2012;**123**:174-181. DOI: 10.1016/j.apcatb.2012.04.009

[27] Zhang YC et al. Novel synthesis and high visible light photocatalytic activity of SnS_2 nanoflakes from $SnCl_2$ $2H_2O$ and S powders. Applied Catalysis B: Environmental. 2010;**95**(1-2):153-159. DOI: 10.1016/j.apcatb.2009.12.022

[28] Zhang Z et al. Ultrathin hexagonal SnS_2 nanosheets coupled with g-C_3N_4 nanosheets as 2D/2D heterojunction photocatalysts toward high photocatalytic activity. Applied Catalysis B: Environmental. 2015;**163**:298-305. DOI: 10.1016/j.apcatb.2014.08.013

[29] Atkin P et al. 2D WS_2/carbon dot hybrids with enhanced photocatalytic activity. Journal of Materials Chemistry A. 2016;**4**(35):13563-13571. DOI: 10.1039/C6TA06415A

[30] Fang HB et al. Efficient charge separation promoting visible-light-driven photocatalytic activity of MnOx decorated WS_2 hybrid nanosheets. Electrochemistry Communications. 2016;**72**:118-121. DOI: 10.1016/j.elecom.2016.09.020

[31] Vattikuti SP, Shim J, Byon C. 1D Bi_2S_3 nanorod/2D e-WS_2 nanosheet heterojunction photocatalyst for enhanced photocatalytic activity. Journal of Solid State Chemistry. 2018;**258**:526-535. DOI: 10.1016/j.jssc.2017.11.017

Chapter 7

Iron Oxide Nanoparticles: An Inorganic Phosphatase

Xiao-Lan Huang

Abstract

Phosphorus is one of the most important macronutrients for the primary production. The transformation of dissolved organic phosphorus in the environment and its contribution to biological production in the different ecosystems is still a mystery. Recently, it was demonstrated that phosphate ester can be rapidly hydrolyzed in solutions containing iron oxide nanoparticles with enzyme kinetics. The catalyst is sensitive to temperature and pH changes and inhibited by tetrahedral oxyanions with an order of $PO_4 < MoO_4 < WO_4$. The oxo-Fe structure in the iron oxide nanoparticles, like the metal center of natural phosphatase (e.g., purple acid phosphatase, PAP), might contribute to the observed catalytic activity. Iron oxide nanoparticles are very common and widely exist in the current earth environment, and phosphate esters are the main component of dissolved organic phosphorus in soil and waters. It is expected that iron oxide nanoparticles in aqueous environments, as an inorganic phosphatase, play a critical role for the phosphorus transformation from the view of the phosphorus cycle.

Keywords: enzyme, hydrolysis, iron oxide, nanoparticles, phosphate ester, phosphorus cycle

1. Introduction

Phosphorus is one of the most important macronutrients for the primary production, which is primarily taken up by plants in the form of phosphate ions (HPO_4^{2-} and $H_2PO_4^-$). Most of the knowledge of phosphorus in the environment, including the phosphorus geochemistry cycle, comes from inorganic phosphates [1–4]. The dissolved organic phosphorus transformation and its contribution to the biological production in the different ecosystems, e.g., soil, lake, estuary and ocean, is still a mystery [5–7]. Recently some limited works have indicated that different phosphate esters, especially monoesters are the main components in the dissolved organic phosphorus in soils [8–11] and waters [12–17], which might be an important source of P phytoavailability and a potential source of water eutrophication. The phosphomonoesters in supra-/macro-molecular structures were found to account for the majority (61–73%) of soil organic P in diverse agricultural soils across the world and the monoester P pool was estimated to account for 33% of the total phosphorus (587 ± 32 kg ha^{-1}) by a recently review [18].

In general, the phosphate ester hydrolysis is catalyzed by various enzymes, including purple acid phosphatases (PAPs), which have been identified and characterized from plant, animal and bacterial organisms [19]. On the other hand,

 IntechOpen

several studies have already demonstrated that the phosphate ester can be hydrolyzed with the interaction of minerals in the aqueous environments [5, 20–25]. Here, the results of laboratory study on the hydrolysis of phosphorus esters, promoted by the iron oxide nanoparticles in water, including the aged nanomolar inorganic iron ion solutions [26–28], were summarized. Additionally, the potential role of inorganic iron oxide nanoparticle for the phosphorus cycles due to the intrinsic phosphoesterase activity is postulated.

2. Promotion effect on the phosphate esters hydrolysis

Usually, phosphate ester in water is quite stable. As an example, hydrolysis Glucose 6-phosphate (G6P), a very common phosphate ester in nature, is a slow process without enzyme in the medium of deionized water (DIW), and becomes even slower in the fresh nanomolar inorganic iron solutions. Inorganic orthophosphate (P_i) in the DIW with the addition of 100 μM G6P at room temperature (22 ± 2°C) was initially 0.90 ± 0.04 μM, which became 4.86 ± 0.26 and 10.35 ± 1.19 μM at 4 and 12 days, respectively. The corresponding P_i in the fresh nanomole inorganic iron solutions (0.5–50 nM $Fe(NO_3)_3$) were 1.35 ± 0.09 and 2.55 ± 0.15 μM.

After G6P was added into an aged 14-month 16.5 nM $Fe(NO_3)_3$ solution (pH 6.30) at room temperature, made by acid-forced hydrolysis [27], the P_i was rapidly released (e.g. the initial 20 μM G6P, as presented in **Figure 1**). Like metal ions as well as natural and biomimetic enzymes, the kinetics of G6P hydrolysis in the aged iron solution can be described as a pseudo-first-order reaction for a fixed concentration of G6P [29–36]. For the initial 20 μM G6P, the decrease in G6P concentration, $[G6P]_t$, due to its hydrolysis can be expressed as a function of hydrolysis time, t, as

$$\log [G6P]_t = -1.31 \times 10^{-5} t - 4.718 \ (r^2 = 0.999) \qquad (1)$$

where $[G6P]_t$ is in M and t is in second.

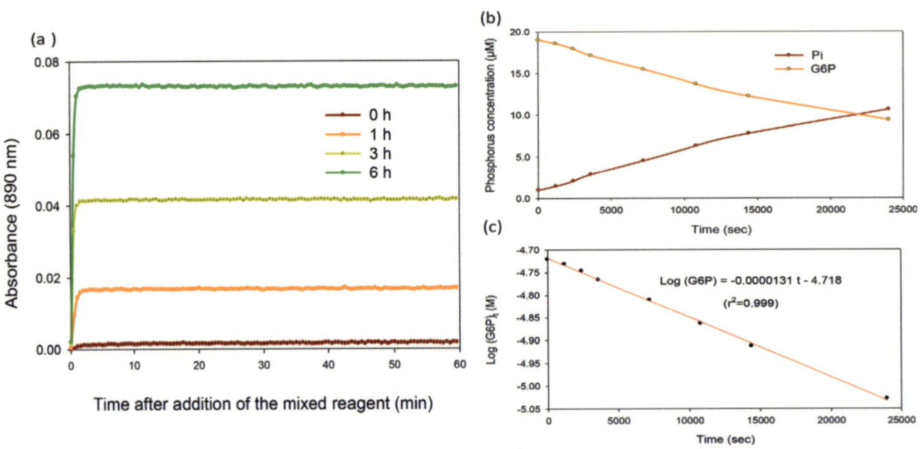

Figure 1.
Hydrolysis of 20 μM G6P in a 16.5 nM $Fe(NO_3)_3$ solution aged 14 months at room temperature (22 ± 2°C). (a) Time courses of formation of phosphorantimonylmolybdenum blue complex from phosphate released from hydrolysis of 20 μM G6P at times of 0, 1, 3, and 6 h, respectively; (b) concentration of P_i and G6P during G6P hydrolysis; and (c) pseudo first-order reaction kinetics of G6P.

The corresponding reaction rate constant (k) was 3.02×10^{-6} s^{-1}, and the half-life ($t_{1/2}$) was 6.38 h. Similar to the initial 20 μM of G6P, the P_i concentration of a initial 100 μM of G6P in the aged inorganic iron solution at 1, 3 and 6.7 h was 4.95, 10.74 and 20.62 μM, respectively. The corresponding k was 8.83×10^{-6} s^{-1}, and the $t_{1/2}$ was 21.8 h. It is highlighted that these k in the aged iron solution were much higher than the previously reported rates in the presence of the fresh unaged nanomolar inorganic iron [26] and millimole metals [30–32] solutions.

Like aged inorganic iron solution, the concentration of phosphate esters and condensed inorganic phosphate decreased, and inorganic orthophosphate (P_i) increased in a solution bearing iron oxide (IO) nanoparticles, which consists of a dialysis membrane tube (DMT, e.g., Spectra/Por 1 membranes, molecular weight cut-off (MWCO) 6000–8000 Da) filled with iron oxide (DMT-IO). The iron oxide (D) was synthesized by $Fe(NO_3)_3$ following the basic protocol of Atkinson [37] and aged at 80°C [28]. The k for 100 μM G6P, Glycerol-2-phosphate (3-carbon, G2P), and three energy metabolism compounds, i.e., adenosine monophosphate (AMP), adenosine diphosphate (ADP), adenosine triphosphate (ATP), as well as two inorganic condensed phosphates, i.e., polyphosphate (poly-P_i), and pyrophosphate (PP_i) at room temperature (22°C) was 2.69×10^{-5}, 1.68×10^{-5}, 1.54×10^{-5}, 5.73×10^{-6}, 5.76×10^{-6}, 3.8×10^{-6}, and 5.09×10^{-6} s^{-1}, respectively. The corresponding $t_{1/2}$ of these phosphorus esters and inorganic condensed phosphates was 7.1, 11.5, 12.5, 33.6, 33.4, 50.7, and 37.8 h, respectively.

Measured k of the initial 20 μM G6P with different sources of iron, either the aged inorganic iron solutions or the solutions bearing inorganic iron oxide nanoparticles (DMT-IO), are listed in **Table 1**. The half-life for aged 4-month Fe $(NO_3)_3$ (16.5 nM), $FeCl_3$ (10 nM) and $Fe(ClO_4)_3$(10 nM) was 37.8, 58.6 and 78.4 h, respectively, whereas the half-life of IO from the same source ($Fe(NO_3)_3$, JT Baker), aged at 5–80°C was 11, 2.7, 3.2 and 2.8 h, though they were in the same order of magnitude. The same patterns were observed for the ATP as well [28]. These results further indicate, as expected, that the behavior of catalysis depends on the sources of iron oxides nanoparticles in solutions—whether $FeCl_3$, or $Fe(NO_3)_3$, and even on the different manufacturers, as well as with the different aging temperatures for IO (5–80°C) [37, 38]. No clear relationships between ferric ion (III) sources, age processing, and catalytic activity, with the hydrolysis rate constant, were observed.

These inorganic iron solutions also have the same promotion effects on hydrolysis of different sugar phosphates, including G2P, ribose-5-phosphate (5-carbon, R5P), and fructose 1-phosphate (6-carbon, F1P) (**Table 2**). As expected, the promotion effect was also found on the hydrolysis of AMP, ADP and ATP, and inorganic condense phosphates (poly-P_i and PP_i) as well as the RNA model compound (4-nitrophenyl phosphate ester, pNPP). However, no promotion effects were observed for the hydrolysis of phosphonates (C-P bonded compounds, e.g., 2-aminoethylphosphonic acid, phosphono-formic acid) and inositol hexakisphosphate (IP6) (data not shown).

As expected, the catalytic activity is related to the soaked time of DIW with DMT-IO and the nature of IO, which can be described by the hydrolysis reaction rate constant. The kinetics of k of 100 μM G6P and ATP in three different IOs is presented in **Figure 2**. These results can be explained by the changes of the nanoparticles concentration in the water. It was expected that the concentration of the IO nanoparticles in these solutions would initially increase up to 10 days and then reach equilibrium. However, the total dissolved iron concentrations in these solutions were still beneath the detection limits of iron (0.1 nM) [39].

Fe source		Manufacturer	Aged time (mo.)	Aged temperature (°C)	IO nanoparticles or total Fe concentration (nM)	20 μM G6P	
						k (10^{-6} s^{-1})	$t_{0.5}$ (h)
Iron oxide nanoparticles (IO)	$Fe(NO_3)_3$	JT Baker	0.25	5	A	16.9	11
	$Fe(NO_3)_3$	JT Baker	0.25	22	B	70.3	2.7
	$Fe(NO_3)_3$	JT Baker	0.25	50	C	59.5	3.2
	$Fe(NO_3)_3$	JT Baker	0.25	80	D	106	1.8
	$FeCl_3$	JT Baker	0.25	25	E	13.8	14
	$Fe(NO_3)_3$	Riedel-de Haën	0.25	50	H	18.5	4.0
	$FeCl_3$	Riedel-de Haën	0.25	50	F	27.4	7.0
	$Fe(NO_3)_3$	Riedel-de Haën	0.25	80	L	80.7	2.4
	$FeCl_3$	Riedel-de Haën	0.25	80	G	54.1	3.6
Aged acidic forced hydrolysis inorganic Fe solution	$Fe(NO_3)_3$	JT Baker	14	22	16.5	30.17	6.4
		Riedel-de Haën	4	22	16.5	5.08	37.8
				22	100	2.96	65.6
			6	22	1000	18.77	10.3
	Iron standard solution (metal Fe in 0.3 M HNO_3)	JT Baker	4	22	1	2.38	81.0
					2.5	4.85	39.7
					7.5	7.22	26.7
					50	6.38	30.2
					100	6.55	29.4
					200	6.58	29.2
					500	8.14	23.6
					1000	5.86	32.8
	$FeCl_3$	JT Baker	16	22	2	6.59	29.2
					10	4.95	38.9
		Riedel-de Haën	4	22	10	3.28	58.6
					100	0.83	231.9
	$FeClO_4$	Aldrich	4	22	10	2.46	78.4
					100	0.61	318.3
	Fe $(NH_4)_2(SO_4)_2$	EM Science	16	22	16.5	9.49	20.3

[a]*The hydrolysis rate constant of 20 μM G6P in the DIW is 1.84 × 10^{-8} s^{-1}, and the corresponding half time is 10,450 h.*

Table 1.
Hydrolysis rate constant of 20 μM G6P in inorganic iron solutions.[a]

3. Kinetics of hydrolysis phosphate esters

As presented in **Table 2**, the hydrolysis reaction rate constant at different initial concentrations of phosphate esters in these aged inorganic iron salt solutions or inorganic iron oxides solutions were not constant. Surprisingly, the *k* from 5 to 250 μM G6P in the 16.5 nM $Fe(NO_3)_3$ solution aged for 14 months at room temperature

Phosphate ester	Fe source	Initial OP (μM)	Rate constant k (10^{-6} s^{-1})	Half-life $t_{0.5}$ (h)
Glycerol-2-phosphate (G2P)	Fe standard solution, 7.5 nM, 4 mo.	10	12.69	15.2
		20	6.4	30.1
		50	3.45	55.8
		500	0.51	374.3
	$Fe(NO_3)_3$, 1000 nM, 6 mo.	20	30.12	6.4
	$FeCl_3$, 2 nM, 16 mo.	20	5.29	36.4
	$Fe(NH_4)_2(SO_4)_2$, 16.5 nM, 16 mo.	20	11.44	16.8
	IO-D (made by $Fe(NO_3)_3$, aged a week at 80°C, soak 1 month at 22°C)	10	85.02	2.26
		50	26.91	7.15
		100	16.39	11.74
Ribose-5-phosphate (R5P)	Fe standard solution, 7.5 nM, 4 mo.	10	13.92	13.8
		20	8.15	23.6
	$Fe(NO_3)_3$, 1000 nM, 6 mo.	20	25.19	7.6
	$FeCl_3$, 2 nM, 16 mo.	20	7.24	26.6
	$Fe(NH_4)_2(SO_4)_2$, 16.5 nM, 16 mo.	20	16.16	11.9
Fuctose-1-phosphate (F1P)	Fe standard solution, 7.5 nM, 4 mo.	10	8.66	22.2
		20	5.25	36.4
	$Fe(NO_3)_3$, 1000 nM, 6 mo.	20	17.08	11.3
	$FeCl_3$, 2 nM, 16 mo.	20	5.29	36.4
	$Fe(NH_4)_2(SO_4)_2$, 16.5 nM, 16 mo.	20	8.5	22.6
Adenosine monophosphate (AMP)	IO-D (made by $Fe(NO_3)_3$, aged a week at 80°C, soak 1 month at 22°C)	10	79	2.44
		100	15.4	12.5
		250	6.2	31.1
Adenosine diphosphate (ADP)	IO-D (made by $Fe(NO_3)_3$, aged a week at 80°C, soak 1 month at 22°C)	10	134	1.44
		100	5.73	33.6
		250	1.59	121
Adenosine triphosphate (ATP)	IO-D (made by $Fe(NO_3)_3$, aged a week at 80°C, soak 1 month at 22°C)	10	61.5	3.13
		25	30.2	6.38
		100	4.4	43.8
	IO-A(made by $Fe(NO_3)_3$, aged a week at 5°C, soak 1 month at 22°C)	20	19.6	9.8
		100	10.8	17.9
	IO-B (made by $Fe(NO_3)_3$, aged a week at 22° C, soak 1 month at 22°C)	20	40.4	4.8
		100	8.46	22.7
	IO-G (made by $FeCl_3$, aged a week at 80°C, soak 1 month at 22°C)	20	38.7	5.0
		100	3.55	54.3
Polyphosphate (poly-P_i)	IO-D (made by $Fe(NO_3)_3$, aged a week at 80°C, soak 1 month at 22°C)	10	67.5	2.85
		100	3.8	50.7
		250	0.39	492
Pyrophosphate (PP_i)	IO-D (made by $Fe(NO_3)_3$, aged a week at 80°C, soak 1 month at 22°C)	10	162	1.19
		100	5.09	37.8
		250	0.71	270

Table 2.
Phosphate ester hydrolysis in the different inorganic iron solutions.

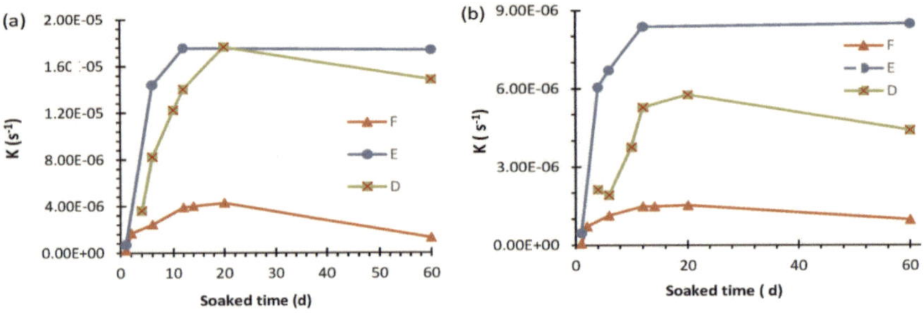

Figure 2.
*Relationship between the soaked time of IO and hydrolysis rate of phosphorus in different DMT-IO solutions: (a) 100 μM G6P, and (b) 100 μM ATP (for details of IOs see **Table 1**).*

can be further described by the Michaelis-Menten equation (**Figure 5a** and **b**), as the typical behavior of biocatalysts. This is contrast to previously reported promotion effects by metals [30–32, 40] and minerals [20, 22, 23, 41, 42]. The maximum k of G6P hydrolysis was about 1 nM s^{-1}, or 3.6 μM h^{-1}, and the Michaelis-Menten constant (K_m) was 13.7 μM of this aged inorganic iron solution.

$$\frac{1}{v} = 9.985 \times 10^8 + \frac{1.371 \times 10^9}{[G6P]_o} \left(r^2 = 0.997\right) \quad (2)$$

In fact, the promotion effect of G6P hydrolysis can be extended to 2500 μM in this aged iron solution with a k of 6.53×10^{-7} s^{-1}, and $t_{1/2}$ of 295 h. It should be pointed out that the concentration of total phosphorus in the solution was 10^3–10^5 higher than that of iron in the solution (e.g., 16.5 nM Fe and 2500 μM G6P).

The same patterns were also observed in the solution bearing inorganic iron oxide nanoparticles. Like the aged inorganic iron solution, the k of various organic phosphate esters or condensed phosphates at different concentrations were not

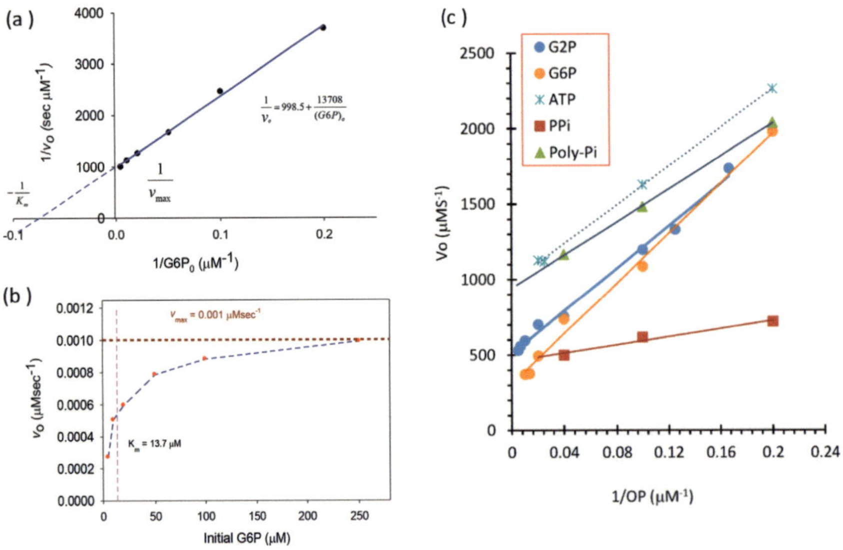

Figure 3.
Kinetics of hydrolysis of phosphate esters in inorganic iron solutions at room temperature (22 ± 2°C). (a) Double reciprocal (initial velocity and initial concentration of G6P) plot of G6P in the 14 month aged 16.5 nM $Fe(NO_3)_3$ solution, (b) initial velocity of G6P hydrolysis (v_o) as a function of the initial concentration of G6P in the aged iron solution, and (c) Lineweaver-Burk plot of different phosphate compounds in a DMT-IO solutions (IO-D).

Phosphorus source	V_m (nM S^{-1})	K_m (μM)	Range (μM)	r^2
G2P	2.0	7.0	6–200	0.99
G6P	3.2	8.3	5–100	0.99
ATP	0.9	9.2	5–50	0.99
Polyphosphate (poly-P_i)	1.1	5.5	5–25	1.00
Pyrophosphate (PP_i)	2.2	1.3	5–25	0.98

Table 3.
Michaelis-Menten constant (K_m) and maximum velocity (V_m) of different phosphorus in a DMT-IO solution (made by $Fe(NO_3)_3{\cdot}9H_2O$ and NaOH, aged a week at 80°C, IO-D, soaked a month).

constant (**Table 2**). The catalytic activity of the different concentration of phosphorus also can be described by the typical Michaelis-Menten equations (**Figure 3c**). Based on the Lineweaver-Burk linear equation ($1/V$ is a linear function of $1/[S]$), the Michaelis-Menten constant (K_m) and maximum velocity (V_m), as well as the range of concentration of phosphorous among these compounds, were determined (**Table 3**). Meanwhile, the catalysis activity was still observed even when the total phosphorus esters exceeded the range of the Michaelis-Menten equations, as with many of the natural enzymes, including the PAP [43].

It should further be pointed out that the similar enzyme kinetics (Michaelis-Menten equations) were observed recently by many inorganic nanoparticles studies, which have been described as nanozyme [44–47]. For example, Fe_3O_4 [44], α-Fe_2O_3 [48], γ-Fe_2O_3 [49], γ-FeOOH [50], Co_3O_4 [51], $MnFe_2O_4$ [52, 53], MFe_2O_4 (M = Mg, Ni, Cu) [54], $ZnFe_2O_4$ [55], NiO [56], and MnO_2 [57] have been observed to have peroxidase-like or catalase-like activity, whereas the vanadium pentoxide (V_2O_5) was demonstrated to have antioxidant enzyme-like (glutathione peroxidase) activity [58–61] and molybdenum trioxide (MoO_3) nanoparticles to have sulfite oxidase activity [62].

4. Inhabitation effects of tetrahedral oxyanions

The hydrolysis of phosphorus ester was significantly inhibited when the tetrahedral oxyanions were introduced into inorganic iron oxides nanoparticle solution, e.g., G6P in a 10-month aged iron solution (**Figure 4**), as the natural PAPs. Both the catalytic and the inhibition behaviors of the catalysis in the presence of 5–125 μM G6P with different tetrahedral oxyanions can be described by a Michaelis-Menten equation (Eqs. (3)–(7)) as follows:

$$\text{without any addition}: \frac{1}{v} = 7.4 \times 10^8 + \frac{2.001 \times 10^9}{[\text{G6P}]_o} \quad (r^2 = 0.945) \tag{3}$$

$$\text{with } 1\,\mu M\ \text{MoO}_4: \frac{1}{v} = 7.4 \times 10^8 + \frac{1.044 \times 10^{10}}{[\text{G6P}]_o} \quad (r^2 = 0.998) \tag{4}$$

$$\text{with } 1\,\mu M\ \text{WO}_4: \frac{1}{v} = 7.4 \times 10^8 + \frac{3.423 \times 10^{10}}{[\text{G6P}]_o} \quad (r^2 = 0.995) \tag{5}$$

$$\text{with } 5\,\mu M\ \text{PO}_4: \frac{1}{v} = 7.4 \times 10^8 + \frac{8.317 \times 10^9}{[\text{G6P}]_o} \quad (r^2 = 0.988) \tag{6}$$

$$\text{with } 10\,\mu M\ \text{PO}_4: \frac{1}{v} = 7.4 \times 10^8 + \frac{1.216 \times 10^{10}}{[\text{G6P}]_o} \quad (r^2 = 0.997) \tag{7}$$

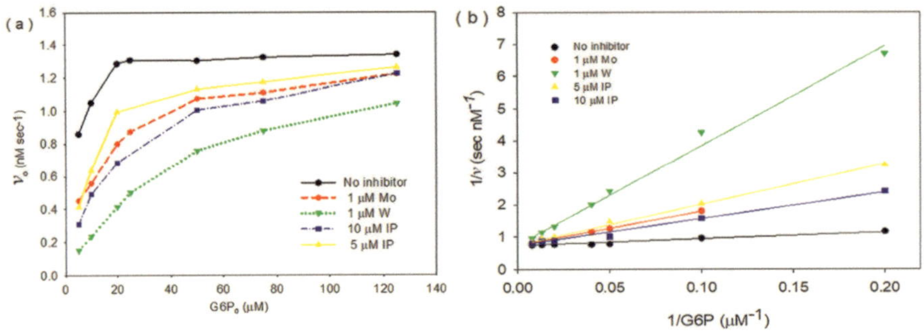

Figure 4.
Inhibiting behavior of different tetrahedral oxyanions on the hydrolysis of Glucose-6-phosphate in an aged 10-month, 1000 nM $Fe(NO_3)_3$ solution at room temperature (22 ± 2°C). (a) Effect of initial concentration of G6P on the initial hydrolysis velocity of G6P, and (b) Lineweaver-Burk plot of aged iron solution in the absence and presence of tetrahedral oxyanions.

The results indicated that the catalysis sites from these catalyst, i.e., the inorganic iron oxide nanoparticles, may be only bound to either the tetrahedral oxyanions (PO_4, MoO_4, and WO_4) or phosphate esters to form an intermediate, but cannot bind both of them at any given moment. The modes of tetrahedral oxyanions and G6P are competitive (**Figure 5**). The K_m, the G6P concentration at which the reaction rate reaches one-half of maximum velocity ($v_{max}/2$), was about 2.7 μM G6P in this aged iron solutions with no inhibitors. The K_{Mapp} with addition 1 μM WO_4, MoO_4, and 5 and 10 μM PO_4 was 46.2, 14.1, 11.1 and 17.1 μM G6P, respectively. Therefore, the K_i of WO_4, MoO_4, and PO_4 are 0.06, 0.24 and 1.6–1.9 μM, respectively. It is interesting to compare the catalytic behavior of these inorganic iron oxides nanoparticles solution to natural PAPs and their biomimetics, though the velocities of hydrolysis G6P in the inorganic catalyst are still lower than that of natural phosphoesterase. For natural PAPs, K_m and K_i of PO_4 is usually in the millimolar range [63–69], only K_i of WO_4 and MoO_4 is in the micromolar range [65, 66, 70–72]. The value of K_m of G6P is 920 μM for PAP extracted from sweet potato [33] and 300–310 μM for those from soybean seed [73]. Besides, the modes

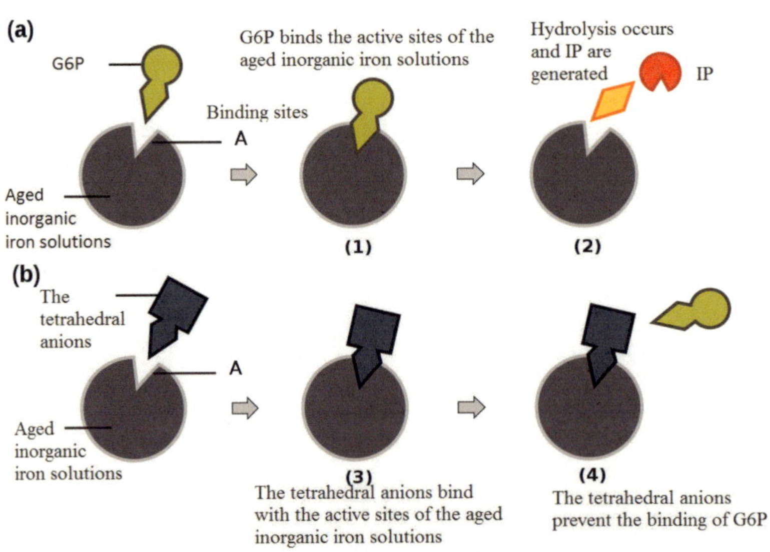

Figure 5.
Diagram of catalysis process of G6P hydrolysis in the presence of the tetrahedral anions in the aged inorganic iron solutions. (a) Reaction without tetrahedral anions; (b) Inhibition with the competitive tetrahedral anions.

of molybdate and tungstate inhibition are noncompetitive [65, 66, 70–74]. Only orthophosphate for natural PAPs are competitive [65, 66] in most cases.

A more significant difference between the inorganic catalyst and the natural phosphoesterase is revealed in their response to the fluoride ion. The activity of all known natural phosphoesterase is very sensitive to fluoride, even at the micromolar level [67, 72–78], while the catalytic activity of the inorganic iron oxides solutions still remain, even when the final concentration of fluoride in the solutions were up to 0.5 M.

5. Effect of temperature

The catalyst on the hydrolysis of phosphate ester is sensitive to temperature, as natural enzymes. The optimum temperature for the phosphate ester hydrolysis reaction by these inorganic catalysts was around 50°C (**Figure 6**), which is comparable to recent observations on the natural enzymes [79–82]. However, catalytic activity of the IO nanoparticles in solution was lost as the temperature was raised to 90°C for an hour or to 72°C for 16 h. This behavior is similar to the thermal denaturation of the natural enzyme. Moreover, the temperature coefficient, Q_{10}, a measure of the hydrolysis velocity, is also decreased as a consequence of increasing the temperature by 10°C. This effect too, is comparable to the general patterns of enzyme behavior in biological systems [80, 81]. Taken together, these observations also carry the implication that moderate, i.e., 50°C, and not high temperatures, were likely favorable to the catalytic reactions from the view of efficiency and speed of the catalyst.

In actuality, the catalytic activity of the nanoparticles remained high after removal from their source (IO) for days, even when stored at −18°C, demonstrated by a storage experiment (**Table 4**) [29], which further suggested that IO nanoparticles can be displaced to a considerable distance from their source and still

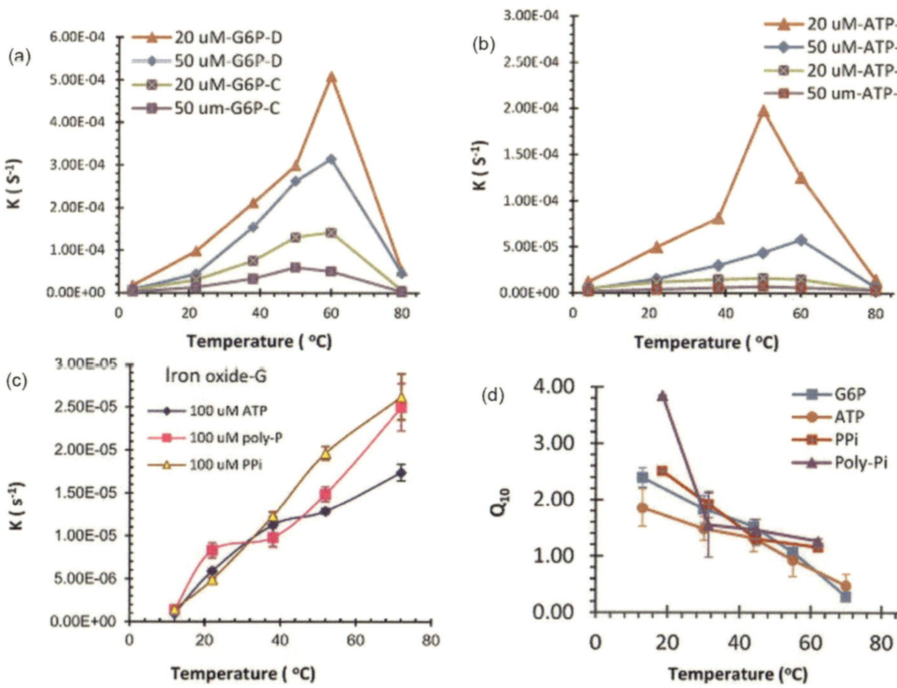

Figure 6.
*Relationship between environment temperature and the hydrolysis rate of phosphorus. (a) G6P, (b) ATP, (c) polyphosphate, and (d) the temperature coefficient Q_{10} (for details of IOs see **Table 1**).*

Treatment		20 μM G6P			20 μM ATP		
		1 h P_i (μM)	5 h P_i (μM)	$t_{0.5}$ (h)	1 h P_i (μM)	5 h P_i (μM)	$t_{0.5}$ (h)
11 days at 22°C	11R	5.49 ± 0.004	13.89 ± 0.064	3.2	4.47 ± 0.016	9.87 ± 0.255	6.0
11 days at 4°C	11 L	5.51 ± 0.010	13.64 ± 0.042	3.3	4.36 ± 0.005	9.71 ± 0.066	6.3
5 days frozen (−18°C) and 6 days at 4°C	5F6L	4.57 ± 0.013	11.48 ± 0.030	4.5	3.76 ± 0.009	7.61 ± 0.043	9.4
5 days at 4°C and 6 days at 22°C	5L6R	5.85 ± 0.011	14.10 ± 0.211	3.1	4.51 ± 0.018	10.30 ± 0.123	5.7
5 days at 50°C and 6 days at 22°C	5H6R	2.93 ± 0.01	6.68 ± 0.020	11	2.71 ± 0.012	4.92 ± 0.032	19
5 days at 50°C and 6 days at 4°C	5H6L	2.62 ± 0.002	5.44 ± 0.001	16	2.48 ± 0.002	3.72 ± 0.04	36
9 days at 50°C and 2 days at 22°C	9H2R	2.43 ± 0.005	4.07 ± 0.005	26	2.38 ± 0.016	3.08 ± 0.025	60

[a]*The P_i of the 20 μM G6P (DIW, control) after 120 h at the room temperature was changed from 1.67 to 1.83 μM. The corresponding half-life was 10,450 h. The P_i of the 20 μM ATP (DIW, control) after 120 h at the room temperature was changed from 1.89 to 2.07 μM. The corresponding half-life was 9290 h.*

Table 4.
Effect of storage conditions on the catalysis activity of iron oxide (IO-F) nanoparticle solutions.[a]

maintain catalytic activities for a considerable time. Meanwhile, low temperatures, even frozen conditions, also favor the persistence of catalytic activity from these IO nanoparticles. These are important from the view of astrobiology (origin of life) [28], but also for plant acquisition, nanoengineering and the potential application for industrial production.

6. Effect of pH and buffer solution

pH is another key factor for enzyme activity. The aged inorganic iron solution or the water bearing iron oxide nanoparticles, e.g., DMT-IO, are generally mildly acidic (pH 5.5–6.5). Various concentrations of bicarbonate were introduced in the DMT-IO system, but in all cases enzyme-like activity for phosphate ester hydrolysis remained quite high (**Figure 7a**). In general, the most favorable pH of the enzyme-like activity was found to be between 6 and 7, though the phosphorus source, the concentrations of bicarbonate, and the type of DMT-IO also influenced its activity (**Figure 7b** and **c**). When pH was raised beyond 7 (e.g., pH 7, 7.2 and 8), the catalysis coefficient, *k*, decreased as the concentration of HCO_3 increased, especially for the DMT-IO-D. When pH in solution was <7 (e.g., pH 6.2, 6.4, and 6.8), however, there were no clear patterns of *k* with respect to the concentration of HCO_3 and both the ATP and G6P in these two nanoparticles-bearing solutions. At the same time, *k* at weak acidic conditions (pH 6.2–6.8) was much higher than at weak base conditions (pH 7–8). This conclusion was further supported by an additional experiment involving the hydrolysis of ATP, whereby the pH values of DMT-IO solution were extended from 4 to 9.3 units by employing four different buffer systems (1.0 M acetate buffer (pH 4.0–5.6), 0.2 M dimethylglutaric acid buffer (pH 4.2–6.8), 20 mM $NaHCO_3$ (pH 7.6–9.3), and 40 mM $NaHCO_3$ (pH 5.8–9.3)) (**Figure 7d**).

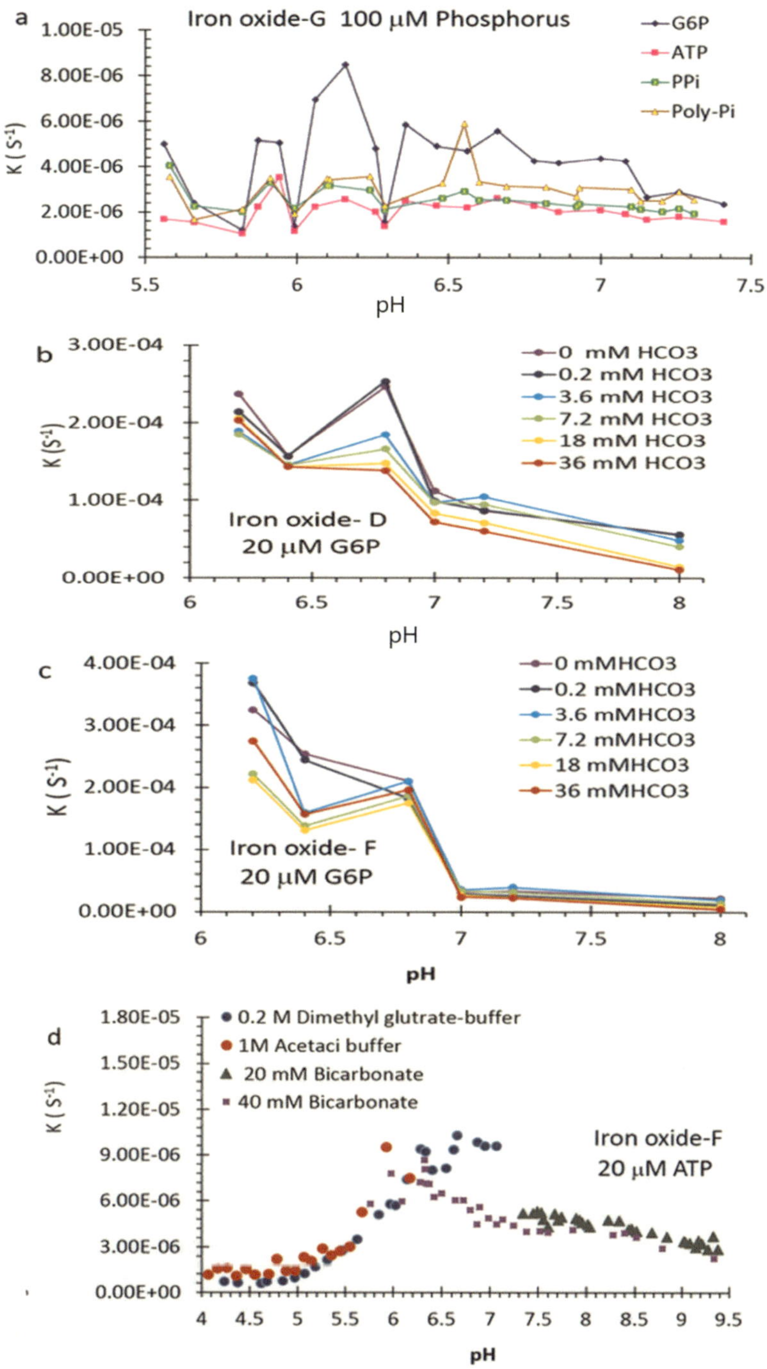

Figure 7.
Relationship between pH and the hydrolysis rate of phosphorus. (a) 100 μM four phosphorus in DMT-IO-G; (b) and (c) bicarbonate concentration on 20 μM G6P DMT-IO-D and F; (d) different buffers for 20 μM ATP in DMT-IO-F (for details of IOs see ***Table 1****).*

It should be pointed out that the catalysis capacity of these solutions bearing inorganic iron oxide nanoparticles is closely related to the buffer used in the system. The catalytic activity dropped precipitously after a small amount of citrate buffer (pH 4.0–6.2) or tris(hydroxymethyl)-aminomethane (TRIS) (pH 5.8–7.2) was introduced into an inorganic iron solution (**Table 5**). Both citrate [83, 84] and

Treatment[a]	Initial G6P (μM)	Reaction time (h)	P_i (μM)	Rate constant k (10^{-6} s^{-1})	Half-life $t_{0.5}$ (h)
DIW and aged 14 mo., 1000 nM, $Fe(NO_3)_3$, (1:1)	20	0	0.22	22.1	8.7
		1	1.75		
		2	3.37		
		6	7.74		
Aged 14 mo., $Fe(NO_3)_3$, 1000 nM and tris–HCl buffer (10 mM, pH 7.0), 1:1	20	0	0.22	1.72	111.8
		1	0.35		
		2	0.41		
		6	0.93		
DIW and aged 14 mo., 1000 nM, $Fe(NO_3)_3$, (1:1)	50	0	0.44	11.3	17.1
		1	2.45		
		2	4.58		
		6	11.14		
Aged 14 mo., $Fe(NO_3)_3$, 1000 nM and tris–HCl buffer (10 mM, pH 7.0), 1:1	50	0	0.44	1.93	99.8
		1	0.71		
		2	0.99		
		6	2.45		

[a]Buffer experiments were conducted with a 1000 nM aged (14-month) inorganic iron solution. The iron solution diluted by DIW (1:1) was used as a control. Either 4 ml of 10 mM Tris-HCl (pH 7.0) or 4 ml of DIW was mixed with a 4 ml of aged iron solution, then 0.1 ml of 1.6 or 4.0 mM G6P stock solution was added to make a final solution containing 20 or 50 μM G6P. P_i concentration was determined at 0, 0.17, 1, 2 and 6 h, respectively, after the addition of G6P. The results of P_i in the table was an average of three replications. The rate constant and half-life were calculated based on the first order kinetics.

Table 5.
Interaction between Tris-HCl buffer and aged inorganic iron solution on the hydrolysis rate of G6P.[a]

TRIS [85] can react with Fe (III) in solution to form the aqueous Fe-complex, particularly at high ratios of citrate or TRIS to Fe (>>10,000:1 molar ratio). But the catalysis of these solutions does not change when the acetic acid-acetate buffer system was introduced, the final acetate concentration in the solution was up to 0.25 M. All of these responses imply that the significant change was not due to the pH itself, but to the interactions between nanoparticles in the solution and chemicals in the environment. It has been reported that some buffer systems can significantly inhibit the activity of natural enzymes, for example, citrate on special PAPs [86] and alkaline phosphatase [87] and TRIS on aminopeptidase and RimO methylthiotransferase [88, 89] due to structure changes and metal-complex formation [84, 85, 90].

7. Natural and inorganic phosphatase

Recall the natural phosphatase, a binuclear metal center (di-iron Fe-Fe or Fe-M (M as Mn and Zn)) that produces orthophosphate due to the net transfer of the phosphoryl group to water, is essential for its catalysis (**Figure 8a**) [19, 66, 91–94]. The μ-(hydr)oxo ligand bridges in the metal center—the key of "phosphoesterase motif"—are a universal feature in binuclear phosphoesterase [19, 94]. They are responsible for the cleavage of phosphoester bonds; including acid and alkaline

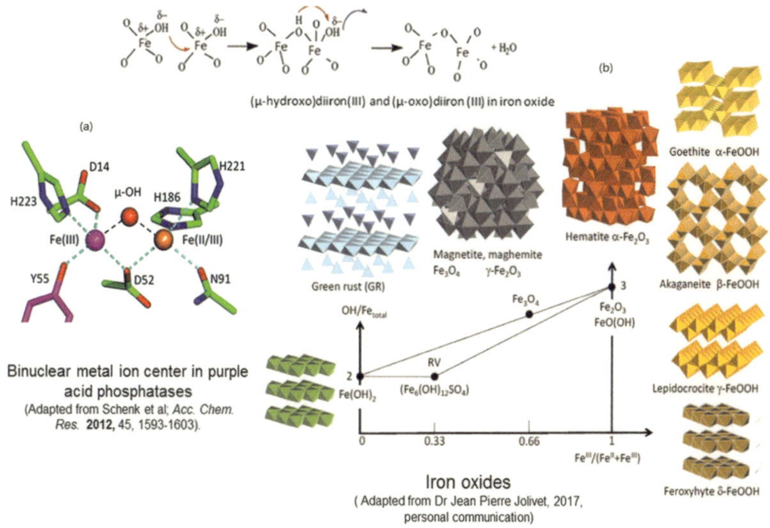

Figure 8.
The metal center of phosphatase. (a) The µ-(hydr)oxo-bridges in purple acid phosphatase (PAP), and (b) Fe-Fe structure in different iron oxides phases.

phosphatases; bacterial exonucleases; diadenosine tetraphosphatase; 5′-nucleotidase; phosphodiesterase; sphingomyelin phosphodiesterase, an enzyme involved in RNA debranching; and a phosphatase in the bacteriophage genome as well as for the family of Ser/Thr protein phosphatase (PP1, PP2A, and calcineurin) [95, 96]. Based on the µ-(hydr)oxo metal bridge structure, different artificial phosphatases have been synthesized by using different organic ligands to stabilize the metal center [19, 97, 98].

It is well known that iron speciation changes due to ion (III) hydrolysis in the solution during the aging process, diiron or polyirons oxide with the oxo-bridge or hydroxo-bridge (bond) might be formed [99–103]. Based on the quantum-chemical calculations by density-functional theory, dihydroxobridging binuclear compounds can be present in aqueous solutions, as binuclear dihydroxobridging $[Fe(H_2O)_4(\mu\text{-}OH)_2Fe(H_2O)_4]^{n+}$ and oxobridging $[Fe(H_2O)_5(\mu\text{-}O)Fe\cdot(H_2O)_5]^{n+}$ (n = 2, 4) cations in the hydrolysis products of cations $[Fe(H_2O)_6]^{m+}$ (m = 2, 3) [104]. The hydroxo-bridged Fe-$(OH)_2$-Fe dimers are the structure units in the polymetric hydroxo complex, which are dependent on pH and aging time [105, 106]. Molecular dynamics simulation further demonstrated the presence of aqueous di-iron or poly-irons, in which the Fe-Fe distance is 3.0–3.5 Å, with bonds by oxo-bridge or hydroxo-bridge [107]. Meanwhile, the (hydr)oxo-bridged Fe-Fe structure has been confirmed by experiments in the interface of iron oxide (IO, solid) to water [108–111]. The µ-oxo iron ion have been identified in-situ in the high concentrated inorganic iron solution (e.g., 0.1 M $Fe(NO_3)_3$ [101], and 0.1 M $FeCl_3$ [112]. The solubility of IOs further indicates that the critical ferrihydrite nucleus with an equivalent diameter of ~15 Å and containing only ~30 Fe atoms is stable in aqueous solution [113]. The 10-angstrom discrete iron-oxo cluster (known as the Keggin ion, Fe_{13}) is also soluble [114], as a constitute structure of ferrihydrite nanoparticles [102]. Consequently, iron oxide nanoparticles with 3.5 Å oxo-Fe bindings (e.g., doubly shared iron octahedra) such as ferrihydrite, goethite, hematite, magnetite, and even green rust (fougerite) can be presented in the natural environment (**Figure 8b**) [115–121]. Therefore, it is reasonable to suggest that the oxo bridged Fe-Fe structure in the aqueous IO nanoparticles contribute to the catalysis of phosphate ester hydrolysis [28]. This compares to the activity of the aged nanomolar inorganic iron ion solutions [26, 27] and mimics of the artificial phosphatases

[19, 97, 98]. The common feature between these IO nanoparticles, either from the DMT-IO or the aged inorganic iron ion solutions, as well as the natural or synthesized biomimetic phosphoesterase, constitute a kind of acceleration of electron transfer rate in the structure of the μ-(hydr)oxo ligand between the metals, particularly iron [19, 26–28, 92, 94, 122, 123]. In other words, the hydrolysis of phosphate ester is entirely dependent on its catalysis on this special Fe-oxo-Fe structure [27, 28]. Experiments and chemical models have also demonstrated that temperature impacts the stability of the aqueous poly-iron formation [124] and the nanostructure of IO in the solution [117, 125], which can explain the thermal denaturation behavior of the inorganic phosphatase (**Figure 6**). The Fe-Fe structure in the nanoparticles due to the nanosize-induced phase transformation and changes in the IO nanoparticle solution with the dissolved CO_2 [126] further supported the response of the inorganic phosphatase at different pH (**Figure 7**).

Similar to phosphatases, the active metal centers of most peroxidase and catalases in nature also comprise the transition metals, for example, horseradish peroxidase, HRP [127], heme catalases [128], uroerythrin [129] with Fe, manganese peroxidase [130], manganese catalases [131, 132] with Mn or haloperoxidases [133] with V, all exhibit the oxo ligand structure. This unique structure might be also accountable for the "intrinsic peroxidase or catalases" from different inorganic metal oxides nanoparticles [44, 48, 49, 51, 53, 54, 58–61, 134, 135]. It was noted that some PAPs were also reported to have activity of peroxidases [136, 137]. The K_m in **Table 3**, which denotes the affinities of the phosphate ester to catalysis, are significantly (up to three orders of magnitude) lower than that of natural PAPs [33, 66, 69, 91, 138, 139]. The same patterns were also observed for these inorganic peroxidases compared to its corresponding HRP [28]. This finding further indicates that the IO nanoparticles are either much more sensitive to the low concentration of phosphate ester or H_2O_2 in the environment, or they have a much higher affinity for phosphate esters or H_2O_2 compared to the natural enzymes, although the maximum velocity of the hydrolysis was relatively low with these IO nanoparticles, especially for high phosphate ester or H_2O_2 concentrations in the environment. Like the nanoparticles of IO and vanadium pentoxide, the intrinsic sulfite oxidase activity of molybdenum trioxide nanoparticles is also due to the oxo ligand of Mo, as revealed in the metal center of sulfite oxidase [62, 140, 141]. Essentially, the catalytic activities depend to some degree on the surface area of these nanoparticles, but not just merely on particle size [48, 51, 52, 134, 142, 143]. The in situ Raman spectroscopy on the changes of V-oxo (V=O) bond in the different V_2O_5 nanomaterials during H_2O_2 catalysis cycle further demonstrated that the catalytic characteristics in these nanoparticles is directly related to the metal structure in the nanoparticle surface [61], which supports the concept of inorganic enzyme [27, 28].

Several recently studies from Europe have suggested that iron-rich nanoparticles (<20 nm) are the main carriers of phosphorus in forest streams and soil solution [10, 11, 144, 145] and monoesters are the main composition of dissolved organic phosphorus in soil and water [10, 15, 16, 18]. This further imply that iron oxide nanoparticles might play a significantly role for the organic phosphorus transformation from the view of phosphorus biogeochemistry, although sorption and precipitation is still the dominant view of the current soil and environmental science on the interaction of iron oxides and dissolved organic phosphorus in soil and sediment [21, 146–149]. A couple of studies still noticed that orthophosphate can be released during the processing of the interactions [5, 20–25]. As iron oxide nanoparticles are very common and widely exist in the soil, sediment, dust, and water [125, 150–153], such enzyme-like catalytic propensities on phosphate esters in the current earth environment may provide an undiscovered feedback of organic phosphorus and play a critical role in the phosphorus cycles.

On the other hand, many effects have been made to improve inorganic nanozyme, both its catalysis capacity and substrate specificity, particularly for the "engineering peroxidase" related to iron oxide for its analytical, biomedical, and environmental applications from the view of nanoengineering [46, 47, 154, 155]. Various polymers or other organic compounds, e.g., porphyrin rings, the backbones of short peptides, amino acids, and even DNA, have been employed in the stabilization of the oxo bridged Fe-metal center in different iron oxides [156–160]. Similar effects should be made for the inorganic phosphatase as well. These "engineering phosphatase" can be employed for environmental monitors after standardization to assess the availability of dissolved organic phosphorus in waters and its potential risk for water eutrophication due to its higher stability and lower cost than protein enzymes, supported by the fact that natural phosphatase has been used for the tool to assess water or soil phosphorus availability [161–164]. Another possibility for industry is to use these high efficiencies engineered phosphatase to release the orthophosphate from the wastewater directly for agriculture.

8. Conclusions and future prospective

Laboratory experiments on the hydrolysis of phosphate ester in water demonstrated that inorganic phosphoesterase-like activity, using various inorganic iron oxide nanoparticles, significantly promotes the hydrolysis of phosphate ester, including G6P, PP_i, and ATP. These findings and the fact that this and other inorganic nanoparticles can act effectively as enzymes: for example, iron oxide as peroxidase, vanadium pentoxide as bromoperoxidase, and molybdenum trioxide nanoparticles as sulfite oxidase; further support the concept of inorganic enzymes. The catalytic property of these nanoparticles is likely due to the structure of the metal oxides or metal bonds in the oxides and not merely to the nanoparticle surfaces. As iron oxide nanoparticles are very common and widely exist in the soil, sediment, and water, such enzyme-like catalytic propensities on phosphate esters, the main composition of dissolved organic phosphorus, in the current earth environment may play a critical role in the phosphorus cycles.

Acknowledgements

This work is dedicated to my beloved parents (Dr. Jing-Xiong Ji and late Dr. Shi-Xiong Huang) and my family (Wei Sun and Jack Jixiang Huang) for their love, endless support, encouragement & sacrifices. X.L.H. greatly appreciates the precious comments from the late Dr. R.J.P. Williams in the past years, and the kindly permissions to use the structure of iron oxide (**Figure 8b**) from Dr. Jean Pierre Jolivet as well as the personal encouragements from Drs. Robert Atlas, Gerhard Schenk, Michael J. Russell, Jia-Zhong Zhang, Raghuraman Venkatapathy and Peter B. Ortner. The experiment part of this work was initially conducted by X.L H from 2007 to 2008 at the AOML, NOAA, supported by the National Oceanic and Atmospheric Administration's (NOAA) Coastal Ocean Program and Climate and Global Change Program. The research was carried out, in part, under the auspices of the Cooperative Institute of Marine and Atmospheric Studies (CIMAS), a joint institute of the University of Miami and NOAA, cooperative agreement #NA67RJ0149. The statements, findings, conclusions, and recommendations are those of the author and do not necessarily reflect the views of CIMAS, NOAA or the U.S. Department of Commerce.

Conflict of interest

The author declares no competing financial interest.

Author details

Xiao-Lan Huang
Independent Researcher, Cincinnati, OH, USA

*Address all correspondence to: xiaolan.huang@ymail.com

References

[1] Paytan A, McLaughlin K. The oceanic phosphorus cycle. Chemical Reviews. 2007;**107**:563-576. DOI: 10.1021/cr0503613

[2] Smil V. Phosphorus in the environment: Natural flows and human interferences. Annual Review of Energy and the Environment. 2000;**25**: 53-88. DOI: 10.1146/annurev.energy.25.1.53

[3] Shen J, Yuan L, Zhang J, Li H, Bai Z, Chen X, et al. Phosphorus dynamics: From soil to plant. Plant Physiology. 2011;**156**:997-1005. DOI: 10.1104/pp.111.175232

[4] Reinhard CT, Planavsky NJ, Gill BC, Ozaki K, Robbins LJ, Lyons TW, et al. Evolution of the global phosphorus cycle. Nature. 2016;**541**:386. DOI: 10.1038/nature20772

[5] Baldwin DS. Organic phosphorus in the aquatic environment. Environmental Chemistry. 2013;**10**: 439-454. DOI: 10.1071/EN13151

[6] Karl DM, Björkman KM. Chapter 5—Dynamics of dissolved organic phosphorus. In: Carlson CA, editor. Biogeochemistry of Marine Dissolved Organic Matter. 2nd ed. Boston: Academic Press; 2015. pp. 233-334. DOI: 10.1016/B978-0-12-405940-5.00005-4

[7] Huang L-M, Jia X-X, Zhang G-L, Shao M-A. Soil organic phosphorus transformation during ecosystem development: A review. Plant and Soil. 2017;**417**:17-42. DOI: 10.1007/s11104-017-3240-y

[8] Dodd RJ, Sharpley AN. Recognizing the role of soil organic phosphorus in soil fertility and water quality. Resources, Conservation and Recycling. 2015;**105**:282-293. DOI: 10.1016/j.resconrec.2015.10.001

[9] McLaren TI, Smernik RJ, McLaughlin MJ, McBeath TM, Kirby JK, Simpson RJ, et al. Complex forms of soil organic phosphorus—A major component of soil phosphorus. Environmental Science & Technology. 2015;**49**:13238-13245. DOI: 10.1021/acs.est.5b02948

[10] Jiang X, Bol R, Cade-Menun BJ, Nischwitz V, Willbold S, Bauke SL, et al. Colloid-bound and dissolved phosphorus species in topsoil water extracts along a grassland transect from Cambisol to Stagnosol. Biogeosciences. 2017;**14**:1153-1164. DOI: 10.5194/bg-14-1153-2017

[11] Missong A, Holzmann S, Bol R, Nischwitz V, Puhlmann H, Wilpert K, et al. Leaching of natural colloids from forest topsoils and their relevance for phosphorus mobility. Science of the Total Environment. 2018;**634**:305-315. DOI: 10.1016/j.scitotenv.2018.03.265

[12] Wang Z-H, Liang Y, Kang W. Utilization of dissolved organic phosphorus by different groups of phytoplankton taxa. Harmful Algae. 2011;**12**:113-118. DOI: 10.1016/j.hal.2011.09.005

[13] Ged EC, Boyer TH. Molecular weight distribution of phosphorus fraction of aquatic dissolved organic matter. Chemosphere. 2013;**91**:921-927. DOI: 10.1016/j.chemosphere.2013.01.113

[14] Ni Z, Wang S, Wang Y. Characteristics of bioavailable organic phosphorus in sediment and its contribution to lake eutrophication in China. Environmental Pollution. 2016;**219**:537-544. DOI: 10.1016/j.envpol.2016.05.087

[15] Bell Douglas W, Pellechia P, Chambers LR, Longo Amelia F, McCabe Kelly M, Ingall Ellery D, et al. Isolation

and molecular characterization of dissolved organic phosphorus using electrodialysis-reverse osmosis and solution 31P-NMR. Limnology and Oceanography: Methods. 2017;**15**: 436-452. DOI: 10.1002/lom3.10171

[16] Duhamel S, Björkman KM, Repeta DJ, Karl DM. Phosphorus dynamics in biogeochemically distinct regions of the southeast subtropical Pacific Ocean. Progress in Oceanography. 2017;**151**: 261-274. DOI: 10.1016/j.pocean.2016.12.007

[17] Liu S, Zhu Y, Wu F, Meng W, Wang H, He Z, et al. Using solid 13C NMR coupled with solution 31P NMR spectroscopy to investigate molecular species and lability of organic carbon and phosphorus from aquatic plants in Tai Lake, China. Environmental Science and Pollution Research. 2017;**24**: 1880-1889. DOI: 10.1007/s11356-016-7954-9

[18] Menezes-Blackburn D, Giles C, Darch T, George TS, Blackwell M, Stutter M, et al. Opportunities for mobilizing recalcitrant phosphorus from agricultural soils: A review. Plant and Soil. 2018;**427**:5-16. DOI: 10.1007/s11104-017-3362-2

[19] Schenk G, Mitić N, Hanson GR, Comba P. Purple acid phosphatase: A journey into the function and mechanism of a colorful enzyme. Coordination Chemistry Reviews. 2013; **257**:473-482. DOI: 10.1016/j.ccr.2012.03.020

[20] Baldwin DS, Beattie JK, Coleman LM, Jones DR. Phosphate ester hydrolysis facilitated by mineral phases. Environmental Science and Technology. 1995;**29**:1706-1709. DOI: 10.1021/es00006a040

[21] Huang X-L, Zhang J-Z. Spatial variation in sediment-water exchange of phosphorus in Florida bay: AMP As a model organic compound. Environmental Science & Technology. 2010;**44**:7790-7795. DOI: 10.1021/es100057r

[22] Olsson R, Giesler R, Loring JS, Persson P. Adsorption, desorption, and surface-promoted hydrolysis of glucose-1-phosphate in aqueous goethite (α-FeOOH) suspensions. Langmuir. 2010;**26**:18760-18770. DOI: 10.1021/la1026152

[23] Persson P, Andersson T, Nelson H, Sjöberg S, Giesler R, Lövgren L. Surface complexes of monomethyl phosphate stabilized by hydrogen bonding on goethite (α-FeOOH) nanoparticles. Journal of Colloid and Interface Science. 2012;**386**:350-358. DOI: 10.1016/j.jcis.2012.07.042

[24] Mäkie P, Westin G, Persson P, Österlund L. Adsorption of trimethyl phosphate on maghemite, hematite, and goethite nanoparticles. The Journal of Physical Chemistry A. 2011;**115**: 8948-8959. DOI: 10.1021/jp201065w

[25] Lü C, Yan D, He J, Zhou B, Li L, Zheng Q. Environmental geochemistry significance of organic phosphorus: An insight from its adsorption on iron oxides. Applied Geochemistry. 2017;**84**: 52-60. DOI: 10.1016/j.apgeochem.2017.05.026

[26] Huang XL, Zhang JZ. Sediment-Water Exchange of Dissolved Organic Phosphorus in Florida Bay. Miami, FL: Cooperative Institute for Marine and Atmospheric Studies; 2007

[27] Huang XL, Zhang JZ. Hydrolysis of glucose-6-phosphate in aged, acid-forced hydrolysed nanomolar inorganic iron solutions—An inorganic biocatalyst? RSC Advances. 2012;**2**: 199-208. DOI: 10.1039/C1RA00353D

[28] Huang XL. Hydrolysis of phosphate esters catalyzed by inorganic iron oxide nanoparticles acting as biocatalysts.

Astrobiology. 2018;**18**:294-310. DOI: 10.1089/ast.2016.1628

[29] Bunton CA, Chaimovich H. The acid-catalyzed hydrolysis of pyrophosphoric acid. Inorganic Chemistry. 1965;**4**:1763-1766

[30] Sigel H, Amsler PE. Hydrolysis of nucleoside phosphates. 6. On the mechanism of the metal ion promoted dephosphorylation of purine nucleoside 5′-triphosphates. Journal of the American Chemical Society. 1976;**98**: 7390-7400

[31] Milburn RM, Gautem-Basek M, Tribolet R, Sigel H. Comparison of the effectiveness of various metal ions on the promoted dephosphorylation of adenosine 5′-triphosphate (ATP) and uridine 5′-triphosphate (UTP)2. Journal of the American Chemical Society. 1985; **107**:3315-3321

[32] Sigel H. Mechanistic aspects of the metal ion promoted hydrolysis of nucleoside 5′-triphosphates (NTPs). Coordination Chemistry Reviews. 1990; **100**:453-539

[33] Schenk G, Ge Y, Carrington LE, Wynne CJ, Searle IR, Carroll BJ, et al. Binuclear metal centers in plant purple acid phosphatases: Fe-Zn in sweet potato and Fe-Zn in soybean. Archives of Biochemistry and Biophysics. 1999;**370**:183-189. DOI: 10.1006/abbi.1999.1407

[34] Blasko A, Bruice TC. Recent studies of nucleophilic, general-acid, and metal ion catalysis of phosphate diester hydrolysis. Accounts of Chemical Research. 1999;**32**:475-484. DOI: 10.1021/ar980060y

[35] Liu C, Yu S, Li D, Liao Z, Sun X, Xu H. DNA hydrolytic cleavage by the diiron(III) complex Fe(DTPB)(μ-O)(μ-Ac)Cl(BF): Comparison with other binuclear transition metal complexes. Inorganic Chemistry. 2002;**41**:913-922. DOI: 10.1021/ic010302h

[36] Karsten P, Neves A, Bortoluzzi AJ, Lanznaster M, Drago V. Synthesis, structure, properties, and phosphatase-like activity of the first heterodinuclear FeMn complex with the unsymmetric ligand HBPBPMP as a model for the PAP in sweet potato. Inorganic Chemistry. 2002;**41**:4624-4626. DOI: 10.1021/ic025674k

[37] Atkinson RJ. Adsorption of potential-determining ions at the ferric oxide-aqueous electrolyte interface. Journal of Physical Chemistry. 1967;**71**: 550-558

[38] Schwertmann U, Cornell RM. Iron Oxides in the Laboratory: Preparation and Characterization. Weinheim, New York, Chichester Brisbane, Singapore, Toronto: WILEY-VCH Verlag GmbH; 2007

[39] Zhang J-Z, Kelble C, Millero FJ. Gas-segmented continuous flow analysis of iron in water with a long liquid waveguide capillary flow cell. Analytica Chimica Acta. 2001;**438**:49-57. DOI: 10.1016/S0003-2670(01)01031-5

[40] Sigel H. Have adenosine 5′-triphosphate ATP4- and related purine-nucleotides played a role in early evolution? ATP, its own ‘enzyme’ in metal ion facilitated hydrolysis! Inorganica Chimica Acta. 1992;**198-200**: 1-11

[41] Torrents A, Stone AT. Hydrolysis of phenyl picolinate at the mineral/water interface. Envirenmental Science and Technology. 1991;**25**:143-149

[42] Baldwin DS, Beattie JK, Coleman LM, Jones DR. Hydrolysis of an organophosphate ester by manganese dioxide. Environmental Science and Technology. 2001;**35**:713-716. DOI: 10.1021/es001309l

[43] Bardsley WG, Leff P, Kavanagh J, Waight RD. Deviations from Michaelis-Menten kinetics. The possibility of complicated curves for simple kinetic schemes and the computer fitting of experimental data for acetylcholinesterase, acid phosphatase, adenosine deaminase, arylsulphatase, benzylamine oxidase, chymotrypsin, fumarase, galactose dehydrogenase, beta-galactosidase, lactate dehydrogenase, peroxidase and xanthine oxidase. Biochemical Journal. 1980;**187**:739-765

[44] Gao L, Zhuang J, Nie L, Zhang J, Zhang Y, Gu N, et al. Intrinsic peroxidase-like activity of ferromagnetic nanoparticles. Nature Nanotechnology. 2007;**2**:577-583. DOI: 10.1038/nnano.2007.260

[45] Lin Y, Ren J, Qu X. Catalytically active nanomaterials: A promising candidate for artificial enzymes. Accounts of Chemical Research. 2014; **47**:1097-1105. DOI: 10.1021/ar400250z

[46] Gao L, Fan K, Yan X. Iron oxide nanozyme: A multifunctional enzyme mimetic for biomedical applications. Theranostics. 2017;**7**:3207-3227. DOI: 10.7150/thno.19738

[47] Karsten K, Nawaz TM, Wolfgang T. A step into the future: Applications of nanoparticle enzyme mimics. Chemistry-A European Journal. 2018; **24**:9703-9713. DOI: 10.1002/chem.201800384

[48] Chaudhari KN, Chaudhari NK, Yu JS. Peroxidase mimic activity of hematite iron oxides (α-Fe_2O_3) with different nanostructures. Catalysis Science and Technology. 2012;**2**: 119-124. DOI: 10.1039/c1cy00124h

[49] Chen Z, Yin JJ, Zhou YT, Zhang Y, Song L, Song M, et al. Dual enzyme-like activities of iron oxide nanoparticles and their implication for diminishing cytotoxicity. ACS Nano. 2012;**6**: 4001-4012. DOI: 10.1021/nn300291r

[50] Peng C, Jiang B, Liu Q, Guo Z, Xu Z, Huang Q, et al. Graphene-templated formation of two-dimensional lepidocrocite nanostructures for high-efficiency catalytic degradation of phenols. Energy and Environmental Science. 2011;**4**:2035-2040. DOI: 10.1039/c0ee00495b

[51] Mu J, Zhang L, Zhao G, Wang Y. The crystal plane effect on the peroxidase-like catalytic properties of Co_3O_4 nanomaterials. Physical Chemistry Chemical Physics. 2014;**16**: 15709-15716. DOI: 10.1039/c4cp01326c

[52] Peng Y, Wang Z, Liu W, Zhang H, Zuo W, Tang H, et al. Size- and shape-dependent peroxidase-like catalytic activity of $MnFe_2O_4$ nanoparticles and their applications in highly efficient colorimetric detection of target cancer cells. Dalton Transactions. 2015;**44**: 12871-12877. DOI: 10.1039/C5DT01585E

[53] Vernekar AA, Das T, Ghosh S, Mugesh G. A remarkably efficient $MnFe_2O_4$-based oxidase nanozyme. Chemistry—An Asian Journal. 2016;**11**: 72-76. DOI: 10.1002/asia.201500942

[54] Su L, Qin W, Zhang H, Rahman ZU, Ren C, Ma S, et al. The peroxidase/catalase-like activities of MFe_2O_4 (M = Mg, Ni, Cu) MNPs and their application in colorimetric biosensing of glucose. Biosensors and Bioelectronics. 2015;**63**:384-391. DOI: 10.1016/j.bios.2014.07.048

[55] Su L, Feng J, Zhou X, Ren C, Li H, Chen X. Colorimetric detection of urine glucose based $ZnFe_2O_4$ magnetic nanoparticles. Analytical Chemistry. 2012;**84**:5753-5758. DOI: 10.1021/ac300939z

[56] Ray C, Dutta S, Sarkar S, Sahoo R, Roy A, Pal T. Intrinsic peroxidase-like activity of mesoporous nickel oxide for

selective cysteine sensing. Journal of Materials Chemistry B. 2014;**2**: 6097-6105. DOI: 10.1039/C4TB00968A

[57] Yang H, Xiong Y, Zhang P, Su L, Ye F. Colorimetric detection of mercury ions using MnO_2 nanorods as enzyme mimics. Analytical Methods. 2015;**7**: 4596-4601. DOI: 10.1039/c5ay00633c

[58] André R, Natálio F, Humanes M, Leppin J, Heinze K, Wever R, et al. V_2O_5 nanowires with an intrinsic peroxidase-like activity. Advanced Functional Materials. 2011;**21**:501-509. DOI: 10.1002/adfm.201001302

[59] Natalio F, André R, Hartog AF, Stoll B, Jochum KP, Wever R, et al. Vanadium pentoxide nanoparticles mimic vanadium haloperoxidases and thwart biofilm formation. Nature Nanotechnology. 2012;7:530-535. DOI: 10.1038/nnano. 2012.91

[60] Vernekar AA, Sinha D, Srivastava S, Paramasivam PU, D'Silva P, Mugesh G. An antioxidant nanozyme that uncovers the cytoprotective potential of vanadia nanowires. Nature Communications. 2014;**5**. Article ID: 5301. DOI: 10.1038/ncomms 6301

[61] Sourav G, Punarbasu R, Naiwrit K, Jemmis ED, Govindasamy M. Nanoisozymes: Crystal-facet-dependent enzyme-mimetic activity of V_2O_5 nanomaterials. Angewandte Chemie International Edition. 2018;**57**: 4510-4515. DOI: 10.1002/anie.2018 00681

[62] Ragg R, Natalio F, Tahir MN, Janssen H, Kashyap A, Strand D, et al. Molybdenum trioxide nanoparticles with intrinsic sulfite oxidase activity. ACS Nano. 2014;**8**:5182-5189. DOI: 10.1021/nn501235j

[63] Burman S, Davis JC, Weber MJ, Averill BA. The interaction of phosphate with the purple acid phosphatase from beef spleen: Evidence that phosphate binding is accompanied by oxidation of the iron chromophore. Biochemical and Biophysical Research Communications. 1986;**136**:490-497

[64] Pyrz JW, Sage JT, Debrunner PG, Que L Jr. The interaction of phosphate with uteroferrin. Characterization of a reduced uteroferrin-phosphate complex. Journal of Biological Chemistry. 1986;**261**: 11015-11020

[65] David SS, Que L Jr. Anion binding to uteroferrin. Evidence for phosphate coordination to the iron(III) ion of the dinuclear active site and interaction with the hydroxo bridge. Journal of the American Chemical Society. 1990;**112**: 6455-6463

[66] Vincent JB, Crowder MW, Averill BA. Spectroscopic and kinetics studies of a high-salt-stabilized form of the purple acid phosphatase from bovine spleen. Biochemistry. 1991;**30**: 3025-3034

[67] Crans DC, Simone CM, Holz RC, Que L Jr. Interaction of porcine uterine fluid purple acid phosphatase with vanadate and vanadyl cation. Biochemistry. 1992;**31**:11731-11739

[68] Schenk G, Carrington LE, Hamilton SE, Jersey JD, Guddat LW. Crystallization and preliminary X-ray diffraction data for a purple acid phosphatase from sweet potato. Acta Crystallographica Section D: Structural Biology. 1999;**55**:2051-2052. DOI: 10.1107/S0907444999012597

[69] Schenk G, Gahan LR, Carrington LE, Mitić N, Valizadeh M, Hamilton SE, et al. Phosphate forms an unusual tripodal complex with the Fe-Mn center of sweet potato purple acid phosphatase. Proceedings of the National Academy of Sciences of the United States of America. 2005;**102**:273-278. DOI: 10.1073/pnas.0407239102

[70] Wang Z, Ming LJ, Que L Jr, Vincent JB, Crowder MW, Averill BA. 1H NMR and NOE studies of the purple acid phosphatases from porcine uterus and bovine spleen. Biochemistry. 1992;**31**: 5263-5268

[71] Lim JS, Aquino MAS, Sykes AG. Mechanistic studies on the reactions of molybdenum(VI), tungsten(VI), vanadium(V), and arsenic(V) tetraoxo anions with the $Fe^{II}Fe^{III}$ form of purple acid phosphatase from porcine uteri (uteroferrin). Inorganic Chemistry. 1996;**35**:614-618. DOI: 10.1021/ic950139o

[72] Elliott TW, Mitić N, Gahan LR, Guddat LW, Schenk G. Inhibition studies of purple acid phosphatases: Implications for the catalytic mechanism. Journal of the Brazilian Chemical Society. 2006;**17**:1558-1565. DOI: 10.1590/S0103-50532006000800011

[73] Ferreira CV, Taga EM, Aoyama H. Glycolytic intermediates as substrates of soybean acid phosphatase isoforms. Plant Science. 1999;**147**: 49-54. DOI: 10.1016/S0168-9452(99)00096-5

[74] Wang X, Que L Jr. Extended X-ray absorption fine structure studies of the anion complexes of FeZn uteroferrin. Biochemistry. 1998;**37**:7813-7821. DOI: 10.1021/bi980150a

[75] Dietrich M, Munstermann D, Suerbaum H, Witzel H. Purple acid phosphatase from bovine spleen. Interactions at the active site in relation to the reaction mechanism. European Journal of Biochemistry. 1991;**199**: 105-113

[76] Crowder MW, Vincent JB, Averill BA. Electron paramagnetic resonance studies on the high-salt form of bovine spleen purple acid phosphatase. Biochemistry. 1992;**31**:9603-9608. DOI: 10.1021/bi00155a012

[77] Reiter NJ, White DJ, Rusnak F. Inhibition of bacteriophage g protein phosphatase by organic and oxoanion inhibitors. Biochemistry. 2002;**41**: 1051-1059. DOI: 10.1021/bi011577b

[78] Balogh E, Todea AM, Muller A, Casey WH. Rates of ligand exchange between $>Fe^{III}-OH_2$ functional groups on a nanometer-sized aqueous cluster and bulk solution. Inorganic Chemistry. 2007;**46**:7087-7092. DOI: 10.1021/ic7009308

[79] Elias M, Wieczorek G, Rosenne S, Tawfik DS. The universality of enzymatic rate-temperature dependency. Trends in Biochemical Sciences. 2014;**39**:1-7. DOI: 10.1016/j.tibs.2013.11.001

[80] Hobbs JK, Jiao W, Easter AD, Parker EJ, Schipper LA, Arcus VL. Change in heat capacity for enzyme catalysis determines temperature dependence of enzyme catalyzed rates. ACS Chemical Biology. 2013;**8**: 2388-2393. DOI: 10.1021/cb4005029

[81] Schipper LA, Hobbs JK, Rutledge S, Arcus VL. Thermodynamic theory explains the temperature optima of soil microbial processes and high Q10 values at low temperatures. Global Change Biology. 2014;**20**:3578-3586. DOI: 10.1111/gcb.12596

[82] Arcus VL, Prentice EJ, Hobbs JK, Mulholland AJ, Van der Kamp MW, Pudney CR, et al. On the temperature dependence of enzyme-catalyzed rates. Biochemistry. 2016;**55**:1681-1688. DOI: 10.1021/acs.biochem.5b01094

[83] Spiro TG, Bates G, Saltman P. The hydrolytic polymerization of ferric citrate. II. The influence of excess citrate. Journal of the American Chemical Society. 1967;**89**:5559-5562

[84] Silva AM, Kong X, Parkin MC, Cammack R, Hider RC. Iron(iii) citrate speciation in aqueous solution. Dalton

Transactions. 2009;**40**:8616-8625. DOI: 10.1039/B910970F

[85] Gupta BS, Taha M, Lee MJ. Stability constants for the equilibrium models of iron(III) with several biological buffers in aqueous solutions. Journal of Solution Chemistry. 2013;**42**:2296-2309. DOI: 10.1007/s10953-013-0107-6

[86] Nuttleman PR, Roberts RM. Transfer of iron from uteroferrin (purple acid phosphatase) to transferrin related to acid phosphatase activity. Journal of Biological Chemistry. 1990; **265**:12192-12199

[87] Evered DF, Steenson TI. Citrate inhibition of alkaline phosphatase. Nature. 1964;**202**:491-492

[88] Desmarais WT, Bienvenue DL, Bzymek KP, Holz RC, Petsko GA, Ringe D. The 1.20 Å resolution crystal structure of the aminopeptidase from *Aeromonas proteolytica* complexed with tris: A tale of buffer inhibition. Structure. 2002;**10**:1063-1072. DOI: 10.1016/S0969-2126(02)00810-9

[89] Molle T, Clémancey M, Latour J-M, Kathirvelu V, Sicoli G, Forouhar F, et al. Unanticipated coordination of tris buffer to the radical SAM cluster of the Rim O methylthiotransferase. JBIC Journal of Biological Inorganic Chemistry. 2016;**21**:549-557. DOI: 10.1007/s00775-016-1365-8

[90] Palmer DA, Drummond SE. Potentiometric determination of the molal formation constants of ferrous acetate complexes in aqueous solutions to high temperatures. Journal of Physical Chemistry. 1988;**92**:6795-6800

[91] LeBansky BR, McKnight TD, Gritting LR. Purification and characterization of a secreted purple phosphatase from soybean suspension cultures. Plant Physiology. 1992;**99**: 391-395. DOI: 10.1104/pp.99.2.391

[92] Sträter N, Lipscomb WN. Two-metal ion catalysis in enzymatic acyl- and phosphoryl-transfer reactions. Angewandte Chemie International Edition. 1996;**35**:2024-2055. DOI: 10.1002/anie.199620241

[93] Lindqvist Y, Johansson E, Kaija H, Vihko P, Schneider G. Three-dimensional structure of a mammalian purple acid phosphatase at 2.2 Å resolution with a μ-(hydr)oxo bridged di-iron center. Journal of Molecular Biology. 1999;**291**:135-147. DOI: 10.1006/jmbi.1999.2962

[94] Mitić N, Smith SJ, Neves A, Guddat LW, Gahan LR, Schenk G. The catalytic mechanisms of binuclear metallohydrolases. Chemical Reviews. 2006;**106**:3338-3363. DOI: 10.1021/cr050318f

[95] Lohse DL, Denu JM, Dixon JE. Insights derived from the structures of the Ser/Thr phosphatases calcineurin and protein phosphatase 1. Structure. 1995;**3**:987-990. DOI: 10.1016/S0969-2126(01)00234-9

[96] Rusnak F, Mertz P. Calcineurin: Form and function. Physiological Reviews. 2000;**80**:1483-1521. DOI: 10.1152/physrev.2000.80.4.1483

[97] Than R, Feldmann AA, Krebs B. Structural and functional studies on model compounds of purple acid phosphatases and catechol oxidases. Coordination Chemistry Reviews. 1999; **182**:211-241. DOI: 10.1016/S0010-8545(98)00234-3

[98] Belle C, Pierre JL. Asymmetry in bridged binuclear metalloenzymes: Lessons for the chemist. European Journal of Inorganic Chemistry. 2003;**23**:4137-4146. DOI: 10.1002/ejic.200300231

[99] Schwertmann U, Friedl J, Stanjek H. From Fe(III) ions to ferrihydrite and then to hematite. Journal of Colloid and

Interface Science. 1999;**209**:215-223. DOI: 10.1006/jcis.1998.5899

[100] Fu D, Keech PG, Sun X, Wren JC. Iron oxyhydroxide nanoparticles formed by forced hydrolysis: Dependence of phase composition on solution concentration. Physical Chemistry Chemical Physics. 2011;**13**: 18523-18529. DOI: 10.1039/C1CP20188C

[101] Zhu M, Puls BW, Frandsen C, Kubicki JD, Zhang H, Waychunas GA. In situ structural characterization of ferric iron dimers in aqueous solutions: Identification of μ-oxo species. Inorganic Chemistry. 2013;**52**: 6788-6797. DOI: 10.1021/ic302053w

[102] Weatherill JS, Morris K, Bots P, Stawski TM, Janssen A, Abrahamsen L, et al. Ferrihydrite formation: The role of Fe13 Keggin clusters. Environmental Science and Technology. 2016;**50**: 9333-9342. DOI: 10.1021/acs. est.6b02481

[103] Das B. Theoretical study of small iron-oxyhydroxide clusters and formation of ferrihydrite. Journal of Physical Chemistry A. 2018;**122**:652-661. DOI: 10.1021/acs.jpca.7b09470

[104] Panina N, Belyaev A, Eremin A, Davidovich P. DFT quantum-chemical study of the hydrolysis products of Fe (II) and Fe(III) aqua-complexes. Russian Journal of General Chemistry. 2010;**80**:889-894. DOI: 10.1134/ s1070363210050038

[105] Pykhteev OY, Efimov AA. Hydrolytic polymerization of iron(III) in partially neutralized nitrate solutions. Russian Journal of Inorganic Chemistry. 1999;**44**:494-499

[106] Baev A, Evsei E. Distribution and reaction dynamics of iron(III) hydrolytic cationic species. Russian Journal of Inorganic Chemistry. 2010;**55**: 508-522. DOI: 10.1134/s00360236 10040054

[107] Zhang H, Waychunas GA, Banfield JF. Molecular dynamics simulation study of the early stages of nucleation of iron oxyhydroxide nanoparticles in aqueous solutions. Journal of Physical Chemistry B. 2015;**119**: 10630-10642. DOI: 10.1021/acs. jpcb.5b03801

[108] Barron V, Torrent J. Surface hydroxyl configuration of various crystal faces of hematite and goethite. Journal of Colloid and Interface Science. 1996;**177**:407-410. DOI: 10.1006/ jcis.1996.0051

[109] Eggleston CM, Stack AG, Rosso KM, Higgins SR, Bice AM, Boese SW, et al. The structure of hematite (a-Fe_2O_3) (001) surfaces in aqueous media: Scanning tunneling microscopy and resonant tunneling calculations of coexisting O and Fe terminations. Geochimica et Cosmochimica Acta. 2003;**67**:985-1000. DOI: 10.1016/ S0016-7037(02)01200-0

[110] Trainor TP, Chaka AM, Eng PJ, Newville M, Waychunas GA, Catalano JG, et al. Structure and reactivity of the hydrated hematite (0001) surface. Surface Science. 2004;**573**:204-224. DOI: 10.1016/j.susc.2004.09.040

[111] Tanwar KS, Lo CS, Eng PJ, Catalano JG, Walko DA, Brown GE Jr, et al. Surface diffraction study of the hydrated hematite surface. Surface Science. 2007;**601**:460-474. DOI: 10.1016/j.susc.2006.10.021

[112] Hellman H, Laitinen RS, Kaila L, Jalonen J, Hietapelto V, Jokela J, et al. Identification of hydrolysis products of $FeCl_3$ $6H_2O$ by ESI-MS. Journal of Mass Spectrometry. 2006;**41**:1421-1429. DOI: 10.1002/jms.1107

[113] Hiemstra T. Formation, stability, and solubility of metal oxide nanoparticles: Surface entropy, enthalpy, and free energy of ferrihydrite. Geochimica et Cosmochimica Acta.

2015;**158**:79-198. DOI: 10.1016/j.gca.2015.02.032

[114] Sadeghi O, Zakharov LN, Nyman M. Aqueous formation and manipulation of the iron-oxo Keggin ion. Science. 2015;**347**:1359-1362. DOI: 10.1126/science.aaa4620

[115] Rose J, Manceau A, Masion A, Bottero JY. Structure and mechanisms of formation of $FeOOH(NO_3)$ oligomers in the early stages of hydrolysis. Langmuir. 1997;**13**:3240-3246. DOI: 10.1021/la962079k

[116] Jolivet JP, Tronc E, Chanéac C. Iron oxides: From molecular clusters to solid. A nice example of chemical versatility. Comptes Rendus-Geoscience. 2006;**338**: 488-497. DOI: 10.1016/j.crte.2006.04.014

[117] Navrotsky A, Mazeina L, Majzlan J. Size-driven structural and thermodynamic complexity in iron oxides. Science. 2008;**319**:1635-1638. DOI: 10.1126/science.1148614

[118] Zegeye A, Bonneville S, Benning LG, Sturm A, Fowle DA, Jones C, et al. Green rust formation controls nutrient availability in a ferruginous water column. Geology. 2012;**40**:599-602. DOI: 10.1130/G32959.1

[119] Michel FM, Ehm L, Antao SM, Lee PL, Chupas PJ, Liu G, et al. The structure of ferrihydrite, a nanocrystalline material. Science. 2007;**316**:1726-1729. DOI: 10.1126/science.1142525

[120] Michel FM, Barrón V, Torrent J, Morales MP, Serna CJ, Boily JF, et al. Ordered ferrimagnetic form of ferrihydrite reveals links among structure, composition, and magnetism. Proceedings of the National Academy of Sciences of the United States of America. 2010;**107**:2787-2792. DOI: 10.1073/pnas.0910170107

[121] Baumgartner J, Faivre D. Iron solubility, colloids and their impact on iron (oxyhydr)oxide formation from solution. Earth-Science Reviews. 2015; **150**:520-530. DOI: 10.1016/j.earscirev.2015.09.003

[122] Lassila JK, Zalatan JG, Herschlag D. Biological phosphoryl-transfer reactions: Understanding mechanism and catalysis. Annual Review of Biochemistry. 2011;**80**:669-702. DOI: 10.1146/annurev-biochem-060409-092741

[123] Duval S, Danyal K, Shaw S, Lytle AK, Dean DR, Hoffman BM, et al. Electron transfer precedes ATP hydrolysis during nitrogenase catalysis. Proceedings of the National Academy of Sciences of the United States of America. 2013;**110**:16414-16419. DOI: 10.1073/pnas.1311218110

[124] Milburn RM, Vosburgh WC. A spectrophotometric study of the hydrolysis of iron (III) ion, II. Polynuclear species. Journal of the American Chemical Society. 1955;**77**: 1352-1355

[125] Guo H, Barnard AS. Naturally occurring iron oxide nanoparticles: Morphology, surface chemistry and environmental stability. Journal of Materials Chemistry A. 2013;**1**:27-42. DOI: 10.1039/C2TA00523A

[126] Chernyshova IV, Ponnurangam S, Somasundaran P. Linking interfacial chemistry of CO_2 to surface structures of hydrated metal oxide nanoparticles: Hematite. Physical Chemistry Chemical Physics. 2013;**15**:6953-6964. DOI: 10.1039/c3cp44264k

[127] Gajhede M, Schuller DJ, Henriksen A, Smith AT, Poulos TL. Crystal structure of horseradish peroxidase C at 215 Å resolution. Nature Structural Biology. 1997;**4**:1032-1038

[128] Bravo J, Verdaguer N, Tormo J, Betzel C, Switala J, Loewen PC, et al.

Crystal structure of catalase HPII from *Escherichia coli*. Structure. 1995;**3**: 491-502. DOI: 10.1016/S0969-2126(01) 00182-4

[129] LeGall J, Prickril BC, Moura I, Xavier AV, Moura JJG, Huynh BH. Isolation and characterization of rubrerythrin, a non-heme iron protein from *Desulfovibrio vulgaris* that contains rubredoxin centers and a hemerythrin-like binuclear iron cluster. Biochemistry. 1988;**27**:1636-1642

[130] Sundaramoorthy M, Kishi K, Gold MH, Poulos TL. The crystal structure of manganese peroxidase from *Phanerochaete chrysosporium* at 2.06-Å resolution. Journal of Biological Chemistry. 1994;**269**:32759-32767

[131] Dismukes GC. Manganese enzymes with binuclear active sites. Chemical Reviews. 1996;**96**:2909-2926. DOI: 10.1021/cr 950053c

[132] Pace RJ, Stranger R, Petrie S. Why nature chose Mn for the water oxidase in photosystem II. Dalton Transactions. 2012;**41**:7179-7189. DOI: 10.1039/ c2dt30185g

[133] Renirie R, Hemrika W, Wever R. Peroxidase and phosphatase activity of active-site mutants of vanadium chloroperoxidase from the fungus *Curvularia inaequalis*: Implications for the catalytic mechanisms. Journal of Biological Chemistry. 2000;**275**: 11650-11657. DOI: 10.1074/ jbc.275.16.11650

[134] Liu S, Lu F, Xing R, Zhu JJ. Structural effects of Fe_3O_4 nanocrystals on peroxidase-like activity. Chemistry (Weinheim an der Bergstrasse, Germany). 2011;**17**:620-625. DOI: 10.1002/chem.201001789

[135] Dong J, Song L, Yin J-J, He W, Wu Y, Gu N, et al. Co_3O_4 nanoparticles with multi-enzyme activities and their application in immunohistochemical assay. ACS Applied Materials & Interfaces. 2014;**6**:1959-1970. DOI: 10.1021/am405009f

[136] Hayman AR, Cox TM. Purple acid phosphatase of the human macrophage and osteoclast. Characterization, molecular properties, and crystallization of the recombinant di-iron-oxo protein secreted by baculovirus-infected insect cells. Journal of Biological Chemistry. 1994;**269**:1294-1300

[137] Veljanovski V, Vanderbeld B, Knowles VL, Snedden WA, Plaxton WC. Biochemical and molecular characterization of AtPAP26, a vacuolar purple acid phosphatase up-regulated in phosphate-deprived Arabidopsis suspension cells and seedlings. Plant Physiology. 2006;**142**:1282-1293. DOI: 10.1104/pp.106.087171

[138] Cashikar AG, Kumaresan R, Madhusudhana Rao N. Biochemical characterization and subcellular localization of the red kidney bean purple acid phosphatase. Plant Physiology. 1997;**114**:907-915. DOI: 10.1104/pp.114.3.907

[139] Bozzo GG, Raghothama KG, Plaxton WC. Structural and kinetic properties of a novel purple acid phosphatase from phosphate-starved tomato (*Lycopersicon esculentum*) cell cultures. Biochemical Journal. 2004;**377**: 419-428. DOI: 10.1042/BJ20030947

[140] Feng C, Tollin G, Enemark JH. Sulfite oxidizing enzymes. Biochimica et Biophysica Acta—Proteins and Proteomics. 2007;**1774**:527-539. DOI: 10.1016/j.bbapap.2007.03.006

[141] Schwarz G, Mendel RR, Ribbe MW. Molybdenum cofactors, enzymes and pathways. Nature. 2009;**460**: 839-847. DOI: 10.1038/nature08302

[142] Wei H, Wang E. Nanomaterials with enzyme-like characteristics

(nanozymes): Next-generation artificial enzymes. Chemical Society Reviews. 2013;**42**:6060-6093. DOI: 10.1039/c3cs35486e

[143] Zhang K, Zuo W, Wang Z, Liu J, Li T, Wang B, et al. A simple route to $CoFe_2O_4$ nanoparticles with shape and size control and their tunable peroxidase-like activity. RSC Advances. 2015;**5**:10632-10640. DOI: 10.1039/C4RA15675G

[144] Baken S, Moens C, van der Grift B, Smolders E. Phosphate binding by natural iron-rich colloids in streams. Water Research. 2016;**98**:326-333. DOI: 10.1016/j.watres.2016.04.032

[145] Gottselig N, Nischwitz V, Meyn T, Amelung W, Bol R, Halle C, et al. Phosphorus binding to nanoparticles and colloids in forest stream waters. Vadose Zone Journal. 2017;**16**. DOI: 10.2136/vzj 2016.07.0064

[146] Leytem AB, Mikkelsen RL, Gilliam JW. Sorption of organic phosphorus compounds in Atlantic Coastal Plain soils. Soil Science. 2002;**167**:652-658. DOI: 10.1097/01.ss.0003034854.98442.39

[147] Celi L, Barberis E. Abiotic stabilization of organic phosphorus in the environment. In: Turner BL, Frossard E, Baldwin DS, editors. Organic Phosphorus in the Environment. Boston: Oxford University Press; 2005. pp. 113-132. DOI: 10.1079/9780851998220.0113

[148] Berg AS, Joern BC. Sorption dynamics of organic and inorganic phosphorus compounds in soil. Journal of Environmental Quality. 2006;**35**:1855-1862. DOI: 10.2134/jeq2005.0420

[149] Ruttenberg KC, Sulak DJ. Sorption and desorption of dissolved organic phosphorus onto iron (oxyhydr)oxides in seawater. Geochimica et Cosmochimica Acta. 2011;**75**:4095-4112. DOI: 10.1016/j.gca.2010.10.033

[150] Wu J, Boyle E, Sunda W, Wen L-S. Soluble and colloidal iron in the oligotrophic North Atlantic and North Pacific. Science. 2001;**293**:847-849. DOI: 10.1126/science.1059251

[151] Li W, Liu D, Wu J, Kim C, Fortner JD. Aqueous aggregation and surface deposition processes of engineered superparamagnetic iron oxide nanoparticles for environmental applications. Environmental Science & Technology. 2014;**48**:11892-11900. DOI: 10.1021/es502174p

[152] Fitzsimmons JN, Bundy RM, Al-Subiai SN, Barbeau KA, Boyle EA. The composition of dissolved iron in the dusty surface ocean: An exploration using size-fractionated iron-binding ligands. Marine Chemistry. 2015;**173**:125-135. DOI: 10.1016/j.marchem.2014.09.002

[153] Colombo C, Di Iorio E, Liu Q, Jiang Z, Barrón V. Iron oxide nanoparticles in soils: Environmental and agronomic importance. Journal of Nanoscience and Nanotechnology. 2017;**17**(7):4449-4460. DOI: 10.1166/jnn.2017.14197

[154] Golchin J, Golchin K, Alidadian N, Ghaderi S, Eslamkhah S, Eslamkhah M, et al. Nanozyme applications in biology and medicine: An overview. Artificial Cells, Nanomedicine, and Biotechnology. 2017;**45**:1069-1076. DOI: 10.1080/21691401.2017.1313268

[155] Zhou Y, Liu B, Yang R, Liu J. Filling in the gaps between nanozymes and enzymes: Challenges and opportunities. Bioconjugate Chemistry. 2017;**28**:2903-2909. DOI: 10.1021/acs.bioconjchem.7b00673

[156] Singh R, Berry RE, Yang F, Zhang H, Walker FA, Ivancich A. Unprecedented peroxidase-like activity of *Rhodnius prolixus* nitrophorin 2:

Identification of the [FeIV=O por•]+ and [FeIV=O por](Tyr38•) intermediates and their role(s) in substrate oxidation. Biochemistry. 2010; **49**:8857-8872. DOI: 10.1021/bi100499a

[157] Wang H, Huang Y. Prussian-blue-modified iron oxide magnetic nanoparticles as effective peroxidase-like catalysts to degrade methylene blue with H2O2. Journal of Hazardous Materials. 2011;**191**:163-169. DOI: 10.1016/j.jhazmat.2011.04.057

[158] Liu B, Liu J. Accelerating peroxidase mimicking nanozymes using DNA. Nanoscale. 2015;**7**:13831-13835. DOI: 10.1039/C5NR04176G

[159] Zhang Z, Zhang X, Liu B, Liu J. Molecular imprinting on inorganic nanozymes for hundred-fold enzyme specificity. Journal of the American Chemical Society. 2017;**139**:5412-5419. DOI: 10.1021/jacs.7b00601

[160] Fan K, Wang H, Xi J, Liu Q, Meng X, Duan D, et al. Optimization of Fe_3O_4 nanozyme activity via single amino acid modification mimicking an enzyme active site. Chemical Communications. 2017;**53**:424-427. DOI: 10.1039/c6cc08542c

[161] Fox TR, Comerford NB. Rhizosphere phosphatase activity and phosphatase hydrolyzable organic phosphorus in two forested spodosols. Soil Biology and Biochemistry. 1992;**24**: 579-583. DOI: 10.1016/0038-0717(92)90083-A

[162] McLatchey GP, Reddy KR. Regulation of organic matter decomposition and nutrient release in a wetland soil. Journal of Environmental Quality. 1998;**27**:1268-1274. DOI: 10.2134/jeq1998.00472425002700050036x

[163] Hoppe HG. Phosphatase activity in the sea. Hydrobiologia. 2003;**493**: 187-200. DOI: 10.1023/A:1025453918247

[164] Turner BL, McKelvie ID, Haygarth PM. Characterisation of water-extractable soil organic phosphorus by phosphatase hydrolysis. Soil Biology and Biochemistry. 2002;**34**:27-35. DOI: 10.1016/S0038-0717(01)00144-4

Chapter 8

Ceria as an Efficient Nanocatalyst for Organic Transformations

Farha Naaz, Umar Farooq and Tokeer Ahmad

Abstract

Valuable chemicals, fuels and pharmaceuticals obtained by the transformation of raw materials have fascinated a lot of researchers in past few decades. However, to reduce problems related to these transformations different green, sustainable and economic techniques have been developed to carry out such organic transformations. Development of nanostructured catalysts has been preferred to accomplish heterogeneous catalytic organic transformations because of greater number of surface-active sites for catalytic processes, high catalyst recovery rate, environment friendly nature and their ease of synthesis. Besides the advances in nanocatalysis, certain challenges including not well-defined morphologies due to loss of control over it and loss of catalytic activity during operation need to be addressed. Ceria is actively investigated in field of catalysis. As a ubiquitous component in catalytic system, its inception is like an irreplaceable component in organic transformations. In this chapter, we appropriately reported various fabricating approaches to synthesize Cerium and CeO_2-rooted nanoparticles and cerium nanoparticles supported on various support materials, accompanied with multimetallic schemes that show notable contribution to the field of catalysis. This comprehensive chapter will provide an improved understanding of nanostructured CeO_2 and will provide deeper insight in the catalysis of Ce-based nanostructured materials and further widen their ambit of applications.

Keywords: ceria, metal oxide, synthesis, nanocatalysis, organic transformations

1. Introduction

Deterioration of environment by excessive dependency on fossil fuel reservoirs encounters an array of challenges for our ecological system. Presently, some other practices such as producing harmful pharmaceuticals, hazardous by-products in manufacturing, etc., results in the severe environmental problems [1]. To encounter these challenges catalysts are needed, but their multiplicity and complexity demand a breakthrough in the approaches in which these catalysts are designed and used [2]. In chemical reactions, catalysts act as unsung heroes which have marked impact on human society [3]. By approaching recent catalytic constituents, moving beyond simple modifications and making efforts to understand the elementary principles, it has been made possible to synthesize and choose suitable catalysts for a provided set of reactants to obtain desired products [4]. These challenges can be resolved by utilizing various techniques like computational modeling, atomic resolution

IntechOpen

microscopy and atomic scale measurements [5]. With the enhancements in chemical sciences, the property which is named as catalysis adorns the chemical reactions in appropriate aspects. Discussions regarding differentiation between homogeneous and heterogeneous catalysts have dominated research in the recent years [6–8]. Therefore, in this chapter we explore novel catalytic systems, which fulfill the specifications of both homogeneous and heterogeneous catalysts like higher activity and better reproducibility.

At this point, nano catalysis comes into the picture and dramatically shaped queries on combining the properties of both the catalytic systems in recent years. As considering nanodimensions, nanocatalysts possess high surface area which provide a better surface to reactants and this property resembles to homogeneous catalysis. The catalyst can act like heterogeneous due to the insolubility in the reaction solvent and hence carried out an adequate separation from the reaction mixture. In view of these significances, nanocatalysts has propelled to the forefront in investigations in recent years. Much attention has been drawn to explore a lot in the field of nanocatalysis and synthesis of nanomaterials for organic transformations. This encourages researchers to develop a simple, efficient, mild, environmentally benign, ligand free, heterogeneous, and reusable nanocatalyst for organic transformations [6]. Prior research has thoroughly investigated nanoscale catalysts in several reactions [9]. It has been recently studied that due to the high surface area and high activity, nanoparticles (NPs) find much importance in catalysis.

Rare earth metals and their complexes exhibit rich variety of solid-state properties and characteristic behavior which make them interesting subjects for catalyzing many organic transformations. There are 14 lanthanides included in the rare earths which include yttrium and scandium also. Among all the rare earths, several researchers corroborate cerium in the field of catalysis. The abundance of cerium in the upper crust is 64 ppm which is relatively higher than copper, tin and any other rare earth elements. Nanoceria has been regarded as the potential catalyst by employing under ligand free conditions in the form of metal, metal oxides for various organic reactions. Cerium due to its outstanding catalytic efficiency and enough abundance makes it useable for a variety of reactions which makes it the selective element for validating the catalytic conversion of the exhaust system of automobiles in automotive industry [10]. Several studies carried out have revealed that due to the high surface area and reactive morphologies of ceria-based nanomaterials can be effectively used as catalysts for organic transformation reactions such as oxidation, reduction, hydrogenation, coupling reactions and many more [11–13] as represented in **Figure 1**. The rare earth elements occur in many minerals inside the earth's crust with quite higher abundance. Being actively investigated, cerium oxide is most imperative and well-known among light rare earth oxides, especially its use in catalysis [14]. Jons Jakob Berzelius and Wilhelm Hisinger were first who discovered elemental cerium in 1803. The exact values of crustal abundance of cerium (average concentration in the earth's crust) are still contentious. For instance, Kleber and Love stated 46 ppm for the cerium crustal abundance in 1963, while Jackson and Christiansen reported the value of 70 ppm in 1993; McGill reported a wide range from 20 to 46 ppm in 1997. Lide reported the value of 66.5 ppm in 1997, representing the intermediate of the various reported values, was commonly accepted for present discussions [15]. Light rare-earth metals occur mainly in the minerals of fluorocarbonate form called bastnasite and phosphate form called monazite. The elemental distribution varies in both minerals and locations. In bastnasite, cerium content is 49.1% with respect to all the rare earth content from Mountain Pass, California, U.S.; while the content is 50.0% located in

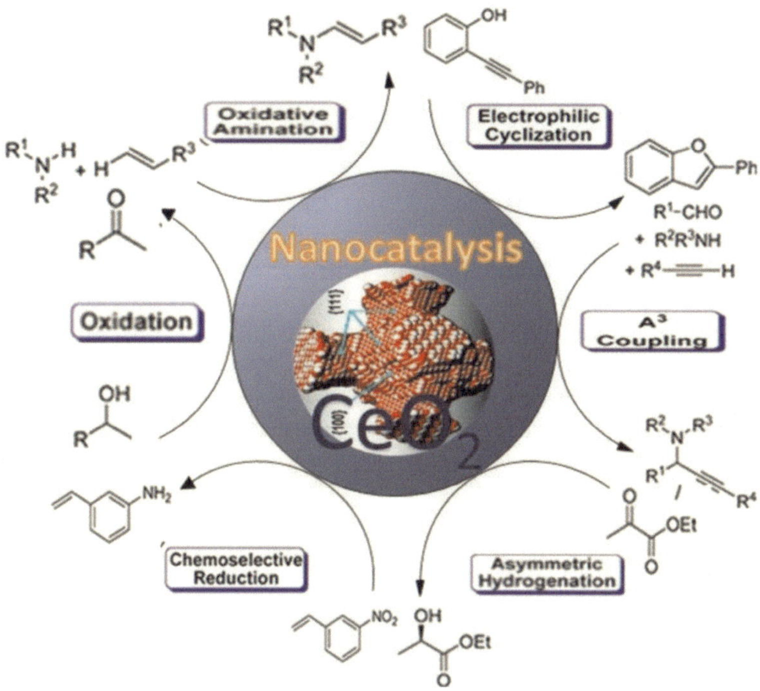

Figure 1.
Nanostructured ceria catalyzed organic transformations.

Bayan Obo, Inner Mongolia, China. The cerium content is 45.8% in monazite minerals at North Staradbroke, Australia and 47% in East Coast Brazil [16]. In the lanthanide group, Cerium is an element with an atomic number of 58. Cerium often shows +3 oxidation state, where it acts as typical rare earth and it also exceptionally has a stable +4 oxidation state.

Metallic cerium, $Ce(OH)_3$ and other oxosalts of cerium like oxalate, nitrate allowed to heat in air or oxygen for the production of its oxide form, i.e., cerium (IV) oxide (CeO_2) [17]. CeO_2 is a well-known cerium compound which is pale yellow/white powder synthesized by cerium oxosalts calcination and generally used as a catalyst or as three-way catalysts (TWCs). The ceria with fluorite structure has tetrahedral holes with oxide ions residing within it, which are formed by the face-centered cubic array of cerium ions, and vacancies are at the octahedral holes. Each cerium ion equivalently surrounded by eight oxygen anions and four cerium cations are tetrahedrally coordinated to these anions. The lattice constant of each unit cell is 5.411 Å. These eight coordination sites are occupied by a cerium cation. It is determined that its structure possess large vacant octahedral holes which intensify its further applications. Ce(III) trioxide (Ce_2O_3) also occurs under ambient conditions [18]. Ce_2O_3 is very unstable against oxidation and as a pressure of 10–40 atm of oxygen is applied, it gets oxidized and then CeO_2 begins to form. Characterization techniques such as X-ray diffraction studied at different temperatures reveals that Cerium oxide have also been observed in other phases. For example, a disordered non-stoichiometric fluorite-related phase of α-phase cerium oxide, is stable above 685°C (CeOx, $1.714 < x < 2$) [19, 20]. A β-phase formed at room temperature with a rhombohedral structure (CeOx, $1.805 < x < 1.812$) remains stable until 400°C [21, 22].

Over the last years, cerium's price has continually dropped as compared to the other rare earth elements. Although fascinating, the cost of cerium oxide cost is going below the price of lanthanum oxide and has observed a sheer incline in its application. CeO_2 has been focused for a plethora of studies both in industry and in academia [23].

2014 and 2015 are the most profused years which have recorded about 2300 publications related to ceria materials. It must be because 1301 publications on catalytic applications in 2015 mark a booming interest in ceria catalysis for the first time become 50% of the total [24–30]. Though, it is worthy to reveal the number of studies on the utilization of ceria in new scientific areas such as biology and pharmaceuticals. Ceria has been used as a support for stem cells cultured in vitro [31] or as a vehicle for intracellular drug delivery [32]. One more noteworthy study reveals that ceria nanoparticles could treat ischemia as well as reduce ischemic brain damage by interruption of the blood-brain barrier after ischemia [33]. Even more fascinatingly, the thermal water and CO_2 splitting by the employment of CeO_2 in solar reactors for fuel generations has been evolving as a novel and exciting investigation topic while accumulation of ceria-based compounds in photocatalysis which merits special mention is another rising field [34, 35].

2. Structure and morphology affecting the catalytic properties of ceria

Structure and morphology play a key role in determining the application of the material by influencing its surface properties. Extensive studies have been conducted to unveil different applications which depend on morphology of nanocrystals. The crystal plane is one of the most common morphological parameters being considered in cerium oxide crystals. The surface of materials is important in various physical and chemical processes that involve the reaction on inorganic oxides such as catalysis and crystallization [29]. CeO_2 in cubic fluorite structure possesses three low-index planes: (100), (110) and (111) as shown in **Figure 2**. The (100) planes contain scattered charged planes which establish a dipole moment perpendicular to the surfaces which are not stable. However, they could be sustained by charge-counteracting species for example, ligands or surfactants or by defects present. The (110) surfaces are charge neutral which consists of anions and cations in stoichiometric proportions in each plane, which exhibit negligible dipole moment perpendicular to the surface. The (111) surfaces also results no dipole moment perpendicular to the surface. Unlike the (110) planes, (111) surfaces consist of a neutral three-plane replicating subdivision ended with a single anion plane. The (100) facet exhibits 2.0 eV of surface energy, the highest one among these three low-index facets and the (111) plane is calculated as the most stable facet irrespective of different potentials used in simulation, both before and after relaxation according to the work done by Vyas. While (110) plane is the other highly stable facet, comprising a surface energy of 1.5 eV from Butler potential calculation [36]. Hence, different shapes of nanostructured ceria particles have different crystal surfaces and plane properties, which further enhances their performances in different systems including catalysis by affecting the interactions between the ceria surface and adsorbed molecules. Fronzi et al. stated similar results on the three low-index surfaces of CeO_2 as they performed density functional theory (DFT) investigations. The stoichiometric (111) surface is the most stable surface structure with a surface free energy of 0.060 eV under oxygen rich conditions calculated by "ab initio atomistic thermodynamics." The subsurface oxygen vacancies of (111) surface has been found to be the most stable one with a surface free energy of -0.001 eV/Å^2 in a reducing environment. While in a highly reducing environment,

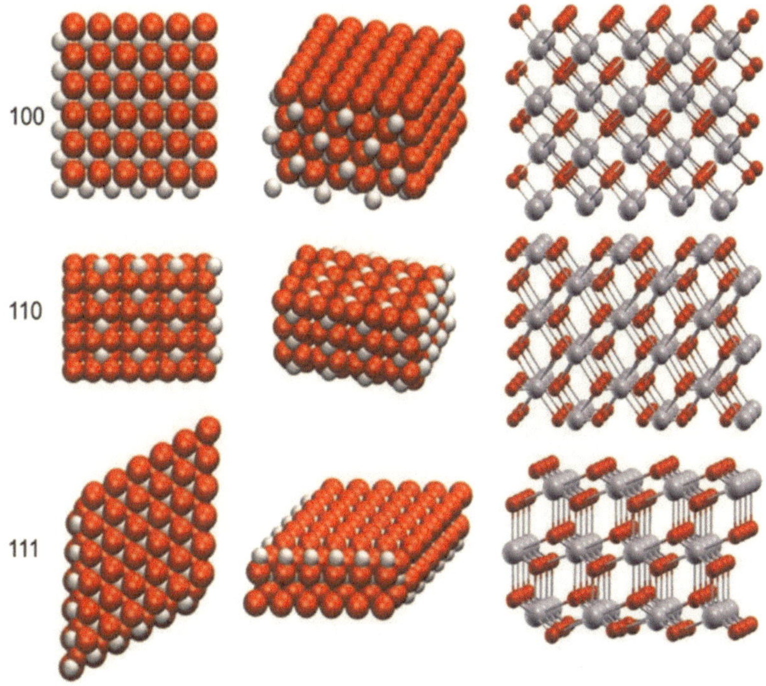

Figure 2.
Diagrammatic representation of CeO_2 facets (100), (110), and (111). Cerium and oxygen ions are represented by red and gray spheres. Reprinted with permission from Ref. [42]. Copyright 2017 American Chemical Society.

a Ce-terminated (111) surface is the most stable one. CeO_2 (110) surface with surface oxygen vacancies has 0.012 eV/Å^2 surface free energy, which is 0.006 eV higher than CeO_2 (111) surface with same oxygen vacancies. The surface free energies of CeO_2 (100) surface having the same type and amount of surface oxygen vacancies terminated with oxygen and cerium are 0.575 and 0.016 eV/Å^2 respectively, which are both larger than those of CeO_2 (111) and CeO_2 (110) surface [37]. Sayle and coworkers reported the surface energies of 11.577 and 2.475 J/m^2 for (331) planes before and after relaxation through applying energy minimization code MIDAS [38]. Other crystal planes of cerium oxides, such as (200), (220), (331) planes, etc., have also been investigated and characterized in both experimental and simulation studies [10, 13]. For example, {220} facets were found in a slightly truncated cerium oxide nanocubes with predominate (100) facets synthesized by Kaneko et al. [39, 40]. Moreover, the feasibility of tailoring the metal oxide morphology have upgraded due to recent advancements in materials chemistry, and the required crystal planes of the cerium oxide materials can be favorably exposed through precise control of the growth kinetics. However, these three low-index planes are the most commonly observed and the most studied facets on synthesized cerium oxide structures [30, 41, 42]. It is also reported that perception about the nanocatalysis must be explained by intrinsic properties of nanoparticles which include (**Figure 3**) (i) quantities such as bond length and binding energy; (ii) quantities related to cohesive energy per discrete atom and the activation energy for atomic dislocation and diffusion, etc.; (iii) properties such as the Hamiltonian which demonstrate band structure, band gap and (iv) properties from the combined effect of binding energy density and atomic cohesive energy like surface area, surface strength, etc. [43].

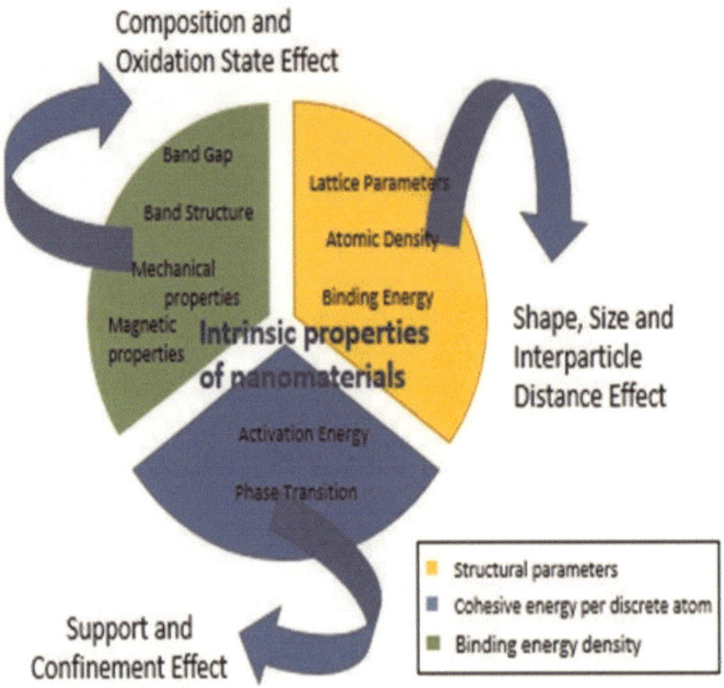

Figure 3.
Intrinsic properties affecting catalytic activity of nanomaterials.

2.1 Mechanism of reactivity and catalytic activity of ceria

In the growing field of catalysis, ceria attributed oxygen storage capacity (OSC) as its fortune. CeO_2 shows multi valence nature which give large number of oxygen vacancies to produce under stoichiometric CeO_2-x at reducing temperatures, which can be oxidized back to CeO_2 in an atmosphere containing oxygen. As cerium (III) (Ce^{3+}) switches to cerium (IV) (Ce^{4+}) states, it results in high oxygen mobility in the ceria lattice that in turn leads to a strong catalytic potential and it happens without any structural modification of the fluorite ceria lattice. The change in energy can cause largely a surface effect to heterogeneous catalyst [44–46].

Thus, scientists have been looking for maximizing the formation of oxygen vacancies of ceria-based catalysts to improve their activity, which needs high temperatures and a reducing atmosphere. Yan's group had made the first observation during the study of preparation of ceria nanostructures, which depict that as compared to octahedral ceria nanoparticles, nanocubes and nanorods had a higher capacity to store and release oxygen at high temperature. Recently, the precise fabrication of functional nanostructured ceria is turn out a routine. Though, some other materials also demonstrate very discrete catalytic activity due to defect sites effect and exposed crystal facet, even from materials which have similar structures [16, 47–51]. The values which are reported in **Table 1** display the exposure of {100} and {110} planes in nano-size ceria accompanying the improved oxygen storage capacity (OSC) and show the following order of OSC nanocubes > nanorods ≫ nanopolyhedra [41]. An appropriate comparison is done with determined hypothetical surface area-normalized OSC which is calculated on more reducible surfaces and specify that OSC is not only defined to the surface, but it also takes place in the bulk [52]. On introduction of defects into the lattice, it is investigated that through controlling synthetic and postproduction parameters, preferred crystal

	OSC[a] (μmol O/g)	OSC/B.E. T (μmol O/m^2)	Calcd OSC[b] (μmol O/m^2)
Nanopolyhedra	318	5.1	6.2
Nanorods	554	9.1	4.9
Nanocubes	353	10.6	5.7

Reprinted with permission from Ref. [41]. Copyright 2005 American Chemical Society.
[a]CO-OSC measured at 400°C.
[b]Calculated according to the theoretical OSC of exposed surfaces. See Ref. [41] for details.

Table 1.
Oxygen storage capacity of variable nanostructures.

orientation is précised. It is also noticed that the reactivity of the surface of crystal is greatly affected by variable lattice defects critically.

Vacancies included in the lattice defects are (a) oxygen vacancy defects, (b) self-interstitials, (c) interstitial impurity atoms and (d) edge dislocations [53]. In the fluorite lattice of ceria, the degree of oxygen mobility accredited to its size, dispersion, and value of oxygen vacancy defects (OVD) [54–57]. The empty 4f states of cerium located electrons by surrounding a vacancy in the ceria support lattice establishes defect sites [17, 58–61]. These defects are shown to be mobile with high oxygen mobility and arise around cerium (III) ions only. On introducing subsurface vacancies into the lattice, the mobility of the vacancies, and therefore the defects, is decreased significantly. The formation of vacancy clusters is in the three or six surrounding cerium ions of the material's surface. An oxygen vacancy defect tends to form under low partial pressure of oxygen [17]. After approaching a favorable oxygen by another oxygen, a bond is formed, and from the surface of the crystal the oxygen molecule can diffuse away. Respectively, the oxygen molecule that is obtained, further diffuses away from the surface and two electrons are left back to be distributed between three cerium atoms. Due to this, cerium atoms undergo partial reduction to a valency between the 3+ and 4+ states. It occurs in a manner to leave behind triads of vacancies which are surrounded by nine cerium atoms sharing eight electrons [62–65]. It is widely proposed that change in the adsorption energy regarding carbon monoxide and oxygen can cause the change in activity of a surface with lattice strain. It is noteworthy that the similar researches reported a lesser increment in the adsorption energy related to carbon dioxide [66, 67]. The adsorption energy for oxygen is about five times superior than for carbon monoxide. Furthermore, the activation barriers respecting to dissociation of oxygen and formation of carbon dioxide are greatly dropped. However, the ease of formation of the oxygen vacancies facilitate the reaction. A molecule of carbon monoxide adsorbs on the surface of the ceria and readily reacts with oxygen existing on its surface and further diffuses away in the form of carbon dioxide leaving an oxygen vacancy. This oxygen vacancy results in a weakened bond between the oxygen atoms, as it allows an adsorbed oxygen molecule to react with the surface. Across the surface of the catalyst, a carbon monoxide diffuses until it encounters the excess oxygen and diffuses away from the surface [68–71].

2.2 Nanoarchitectured ceria and its influence in the behavior to supported metals

The vast expansion of the usage of the nanomaterials offer is just incredible. Nanotechnology revolution has revolutionized the research arena as matter of the fact that it creates the vast possibilities to fabricate the materials with nanodimensions. Catalysis, fuels and microelectronics are different fields of applications where nanostructured cerium oxides grow rapidly and reflecting their importance in enhancing the performances of those systems. Variable morphologies of ceria nanoconstructs have been explored in these applications, for instances nanocubes, nanorods, octahedron polyhedron, tube and many more. Ceria nanostructures with various shapes possess the different crystal planes and surface morphologies, which influences the interactions between the ceria surface and adsorbed molecules, and hence changes the performances in different systems. Zhou et al. described many strategies for synthesis of well-controlled morphologies of nanostructured ceria. Now, Ce-based materials with controlled morphologies which exhibit zero-, one-, two-, and three-dimensional structures are possibly synthesized. (**Figure 4**). The categories defined on basis of number of dimensions which cannot be restricted to the nano-range (<100 nm).

Nanostructured ceria with Zero-dimensional (0D) possessing isotropic cubic phase of the fluorite structure can be observed distinctively. According to results, it presented a lack of fortunate growth direction of seeding crystals. Hence, (0D) nanostructures have most straightforward synthesis. Mono-dimensional (1D) CeO_2 nanoparticles possess the different properties due to which they have been explored more than the 2D and 3D architectures and a variety of synthesis procedures were proposed. 1D hexagonal Ce_2O nano-rods (NRs) synthesized by template-free electrochemical growth method on a Ti substrate which mainly exposes the {110} planes and displayed outstanding photocatalytic activity in hydrogen evolution, with H_2 yield reaching 741 mmol g^{-1} [26].

The preparation of 2D and 3D architectures are drawing significant attention and they also evolving as good alternatives in various catalytic and energy applications. For the construction of a spongy mesoporous CeO_2 microspheres an analogous concept was assumed in which in-situ formation of the removable template by graft polymerization reaction between acrylamide and glucose takes place [72]. Ceria nanocubes synthesis is important because their possession of high surface energies usually exhibit specific activities due to the unsaturated coordination atoms, atomic steps and ledges [73–75]. Numerous fabrication methods have been reported for spherical nanostructures [76–79]. Planes in ceria octahedron have gained much attention as these planes are exposed on their surfaces [80]. Nanostructured ceria with different morphologies, such as nanotubes, spindles, nanosheets, etc. have been synthesized [45, 80, 81]. Due to their enormously developed activities ceria nanorods have increased wide-ranging interest than those of ceria with other shapes in many different reactions, such as CO oxidation, NO reductions and 1,2-dichloroethane and ethyl acetate oxidation [57, 82, 83]. By means of the most stable (111) planes on the surface, ceria octahedra demonstrated the least catalytic activity being studied when compared to the activities of nanocubes, nanorods and other shapes in many reactions, such as CO oxidation and ethyl acetate oxidation [80, 83], Notable progress has been made to achieve these

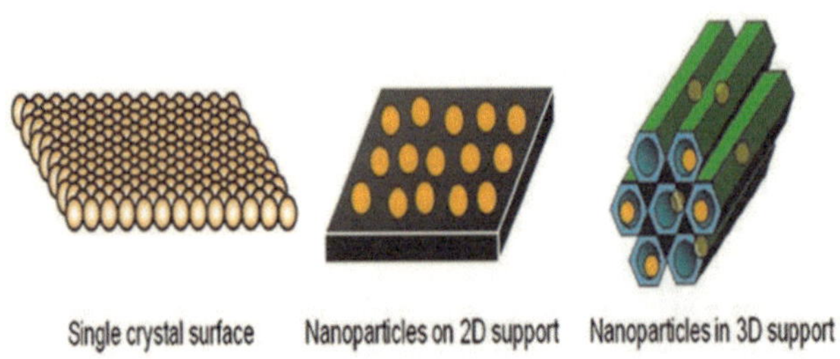

Figure 4.
1D, 2D and 3D nanostructures. Reprinted with permission from Ref. [2]. Copyright 2010 Springer Nature.

ceria nanomaterials. However, synthesizing these morphologically different nanostructures with well-controlled size and homogeneity is still difficult due to their uncommon shapes. For instance, it is difficult to prepare nanosheets due to their exceptionally small thickness and possible quantum size effects [84].

In Prospect, the synthesis of ceria NPs requires the interaction with metal nanoparticles that act as the "active sites" for catalysis. Schelter and co-workers reported a ligand to vary the stability of the Ce (III)/Ce (IV) redox couple, in their synthesized 1,3-bis [(20 tertbutyl) hydroxyamino phenyl]-benzene Ce complex, [85]. Recently, a hydrothermal process has been used for the synthesis of Au@CeO_2, presenting core-shell systems grounded on other precious metal core-shell [86], while Ag@CeO_2 was also synthesized by reverse micelle/redox reaction [86–90]. Among different characterization, an exclusive strength of the STM technique is the ability to enquire the atomic structure of surfaces, down to the level of distinct defects and adsorbates. **Figure 5** shows one such image, obtained on the surface of a CeO_2 (111)/Pt (111) system [91].

Catalytic characteristics of supported metal nanoparticles depend on the role of the support as well as on the composition, shape, particle size, and chemical state too. The catalytic reactivity is directly related to the atomic interaction within support and metal nanoparticles which is termed as metal-support interaction which has attained significant attention nowadays (**Figure 6**). Due to the possession of unique properties by the ceria by virtue of which it makes oxygen species readily available to the metal site which make its outstanding applications in large number of catalytic reactions. This way, noble metals on ceria are activated for various oxidation reactions at low temperatures [92].

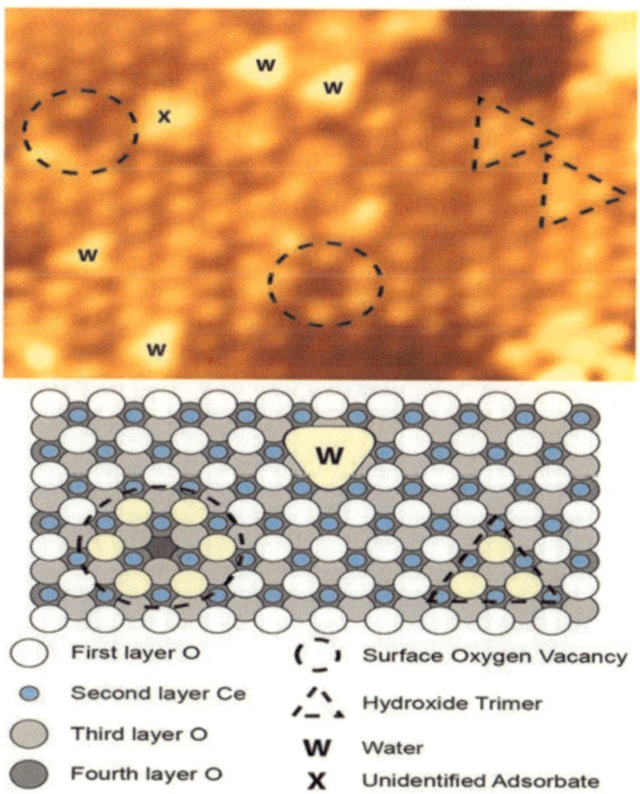

Figure 5.
STM image of the CeO_2 (111)/Pt (111). Reprinted with permission from Ref. [91]. Copyright 2010 American Chemical Society.

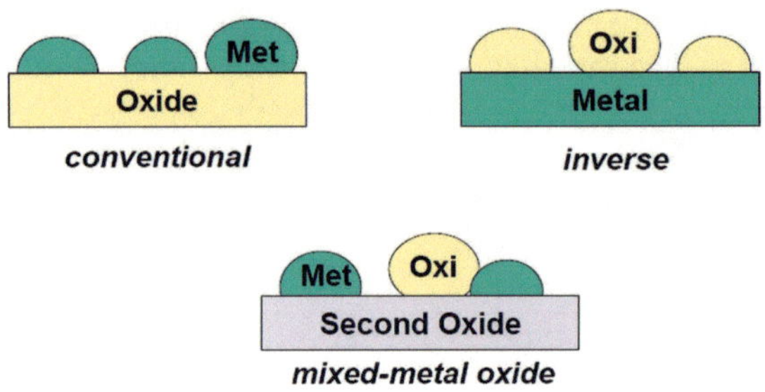

Figure 6.
Metal-oxide configurations used in studies with ceria model catalysts. Reprinted with permission from Ref. [92]. Copyright 2017 Royal Society of Chemistry.

A pivotal role is played in the activity of catalysts by the nature of their support. Vayssilov et al. has studied the origin of interactions proposed between the various support effects like the active (metal) phase and support which include interaction of electrons among both components [93], destabilization or stabilization of particle sizes or shapes [94], surface transport of adsorbates through the boundary (spillover, reverse spillover capture zone effects); [95] and the stabilization/destabilization of oxidized active phases by the support or strong "metal-support interactions" relating movement of partially reduced oxides onto the active phase [96–98]. The metal oxide intervenes in the catalytic process as well as an inert support [99].

Pure ceria, CeO_2, undergo degradation with time at elevated temperatures which minimize its performance due to reduction in its surface area as well as oxygen storage capacity (OSC), also it has been presented that pure ceria accommodate "active" weakly bound oxygen species, which relates bulk rather than to the surface by using steady-state CO oxidation kinetics and/or temperature-programmed desorption (TPD) [100–102]. Thus, development of CeO_2 based nanocatalysts for chosen activities is surely done with a keen understanding about metal-ceria support interaction in supported metal catalysts (**Figure** 7). E. Mamontov used pulsed neutron diffraction to investigate the nature of these "active" oxygen species in pure ceria. The study of oxygen position in oxides by neutron diffraction demonstrates a comparable scattering contrast of oxygen and metal ions. In the real space the oxygen defects in CeO_2 examined by both pulsed neutron diffraction data and atomic pair-distribution function (PDF) analysis whereas in the reciprocal space, it is analyzed by the Rietveld refinement [103].

Zirconia, ZrO_2, has been actively investigated in many studies and have been characterized. The enhanced OSC of ceria-zirconia related to ceria as well as known to improve partial degradation of ceria at high temperatures. It is also probable that

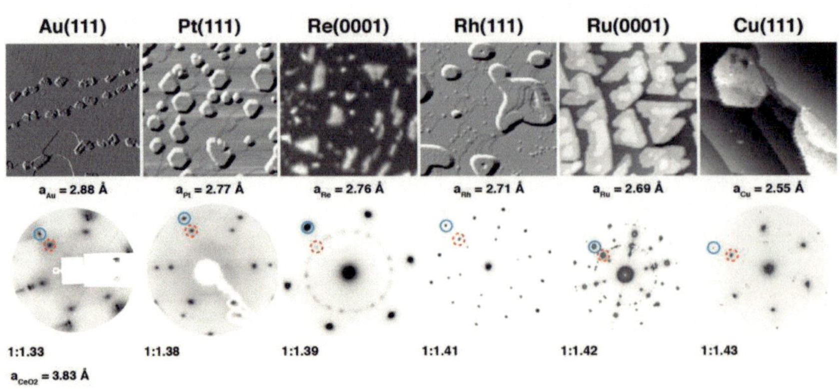

Figure 7.
Real (STM, LEEM) and reciprocal space (LEED) of ceria structures on a range of metals. Reprinted with permission from Ref. [112]. Copyright 2016 American Chemical Society.

the necessary oxygen mobility which is essential for the functioning of CeO_2 as a catalytic support is provided by the interstitial oxygen ions which act as "active" sites [104–111].

CeO_2 impart high oxygen storage and release capacity and it is predicted as the finest supporting material for catalysis at Au NPs following other properties such as facile oxygen vacancy formation, and narrow Ce f-band. Au gets oxidized once in contact with CeO_2. The atomic and electronic interaction between reduced CeO_2 and supported Au NPs is highly contributed by the electrons located on the occupied 4f-orbital of Ce^{3+} ions. Ceria particles were also deposited on the surfaces of Au, Pt, Re, Rh, Ru, Cu to produce inverse oxide/metal catalysts and these supports can cause dramatic effect on the structure of ceria islands in an inverse catalyst [112, 113]. The electron transfer is induced from metal to the support because of the presence of platinum over ceria nanoparticles leading to the formation of a small fraction of Ce^{3+} cations. Among transition metal core-Pt shell nanoparticles, altering Au NPs with alloying elements would be an interesting strategy for lowering CO adsorption energy, as well as deducing and enhancing saturated CO and O_2 concentration, correspondingly. Vayssilov et al. studied that a crystalline atomic arrangement exhibited by the interaction of an illustrative metal cluster Pt8 with two group of model ceria nanostructures [114]. A Pt8 cluster was selected as a model which supported on a stoichiometric $Ce_{40}O_{80}$ nanoparticle. The metal group relates five Pt atoms with ceria particle found as most stable structure for Pt8/ $Ce_{40}O_{80}$. Liu and colleagues [115] prophesied adsorption and dissociation of oxygen and transport processes on the two most stable Ag (111) and Ag (110) surfaces and reveals a binding energy of CeO_2 catalysts and on a monolayer silver supported by CeO_2 (111) surfaces with or without oxygen vacancies by DFT with PAW method. The computed energies of these reactions display that the process of oxygen reduction and the combination of the dissociated oxygen ions in the oxide electrolyte prefer taking place in the triple phase boundaries (TPB) region with oxygen vacancies [116].

2.3 Synthesis and characterization of ceria nanoparticles

This segment of the chapter encompasses the detailed outline of various synthesis techniques and the conditions applied for the reaction which influences the final product. According to the applications in various fields, the synthesis of ceria nanoparticles with desired morphology is very important; therefore, thorough investigations were done by researchers to investigate several approaches. The studies on Ceria based nanoparticles demonstrated that conditional to the synthesis methodology, wide variations may occur in shape, size, crystal structure, and properties of nanostructures, as well as the physical and chemical conditions employed during the reaction process. Current literature revealed many chemical methods, such as, hydrothermal method, co-precipitation method, a micro emulsion mediated approach, and other methods like sol-gel synthesis have been employed to synthesize cerium based nanostructured materials.

Hydrothermal method refers to the oxide synthesis and crystal growth in aqueous solutions under high temperature and pressure using a sealed heated vessel which is known as autoclave. It is well-established method for the laboratory and industrial scale synthesis of nanoceria materials. Two big advantages of this method are that: the reaction temperature is below the melting point of reactants, and the operational parameters such as reaction temperature, duration, autoclave types can be easily tuned to modify the reactivity of synthesized inorganic solids [117]. A facile hydrothermal method for the synthesis of ceria nanocubes with six {100} facets using oleic acid as the surfactant was developed by Wang and co-workers. The as-prepared ceria was single crystalline, confirmed by uniform crystal lattice fringes. The sizes were well controlled with side lengths from 9 to 17 nm [92].

A single-step hydrothermal method is utilized in the fabrication of uniform CeO_2 nanoparticles with diameter approving nano-size, i.e., 13–17 nm and further, hexamethylenetetramine is added for the formation of (220)-dominated surface structure.

Co-precipitation synthetic method is another extensively used approach for preparing nanomaterial. The strategy is best choice for commercial synthesis of CeO_2 and due to very low solubility of ceria, it attains great advantage. This method is simple and rapid preparation process which makes it easy to synthesize controllable particle size and flexible in altering overall homogeneity of the particle with its surface state [118]. The cerium precursors are generally inorganic cerium salt, such as $Ce(NO_3)_3$, $CeCl_3$, $(NH_4)_2Ce\ (NO_3)_6$, and the precipitating agents are usually NaOH, NH_4OH, hydrazine and oxalic acid [119–122]. Abimanyu and coworkers performed co-precipitation method and applied ionic liquid as a template to prepare magnesium and cerium mixed oxides. To overcome the difficulty of controlling particle size, template-assisted co-precipitation, carbonate co-precipitation, redox co-precipitation, etc. have been presented in conventional co-precipitation method [118].

Sol-gel method is highly suitable for the fabrication of metal oxides (**Figure 8**). This method is widely used in ceramics industry and materials science for producing solid materials such as ceramic fibers and dense films. It is easy to accomplish and does not need any special conditions and equipment [117]. The process involves conversion of metal alkoxide/chloride solution into a colloidal suspension (sol) and gelation of the sol to form discrete particles or network polymers in a continuous liquid phase (gel) [123]. Gnanam et al. successfully prepared nanocrystalline cubic fluorite/bixbyite CeO_2 or α-Mn_2O_3 via simple sol-gel method using cerium (III) chloride/manganese (II) chloride as the precursor by using methanol as a solvent calcined at 400°C [124].

The reverse micelle technique offers the greatest control over size and morphology. It is a wet chemical method in which pools of water are enclosed by surfactant molecules in an excess volume of oil. During synthesis procedure, surfactant molecules retain particles separated and confine particle growth this keeps control on size and shape of particles. The root for the technique is the use of a surfactant to stabilize variable aqueous droplet sizes in hydrocarbon medium. Metal salt precursors are transformed by a reactant from the hydrocarbon phase and are contained in

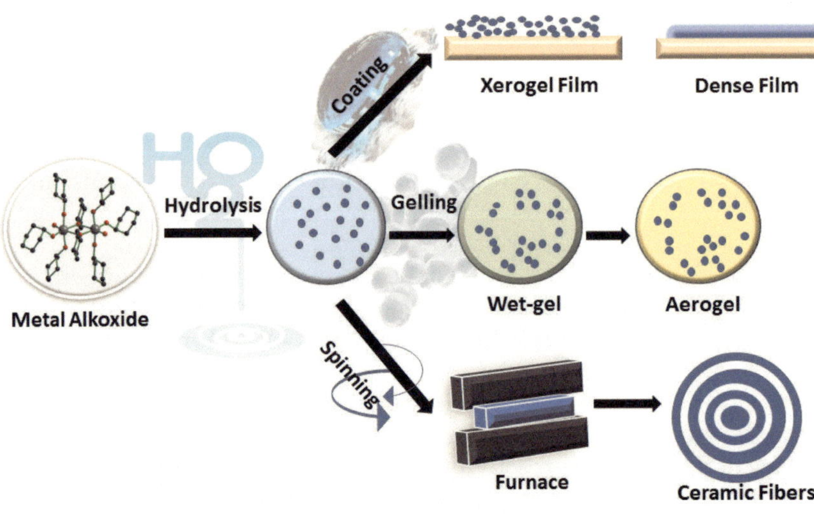

Figure 8.
Diagrammatic illustration of different stages of sol-gel method. Reprinted with permission from Ref. [118].

the aqueous portion. Mutually, the structure of the surfactant and the steric size can produce metals having a wide range of grain sizes. Masui et al. [125] synthesized ceria nanoparticles by using reverse micelles and reported fluctuation of bandgap values from 3.38 to 3.44. Ganguli et al. fabricated monophasic nanoshaped oxides by performing similar versatile methodology [126, 127].

The worth of morphology-activity association is clearly recognized with the implementation of different examples as breakthrough. With the advanced characterization techniques included SEM, HR-TEM, STM, uncountable studies investigated that morphology/exposed facet combination still contain some uncertainties and, so, on the mechanism of crystal growth. Some focus is also bounded by surface reactivity analysis or through TEM. The elucidation of this area is must to simplify instrumentation acquaintance [128, 129]. The powder XRD patterns of the CeO_2 nanorods, nanocubes and nanopolyhedra are of pure cubic phase shows its fluorite structure with lattice constants of 5.414(3), 5.436(3), and 5.405(3) Å, respectively (**Figure 9a**). The enlargement of the reflections indicated their nanocrystalline nature, recognized to the polyhedron and rods distinctly. The sharper reflections for cubes implied their larger sizes as compared with the former two samples. X-ray photoelectron spectroscopy (XPS) and X-ray absorption near edge spectroscopy (XANES) techniques investigate the oxidation state of cerium ions in ceria nanoparticles. The remark of the existence of the Ce^{3+}/Ce^{4+} shifts in150 mm active region leads to the inference that the lateral electron transport and surface reaction kinetics on the thin ceria electrodes are co-limiting processes. The XPS spectrum of the CeO_2 nanorods is shown in **Figure 9b**. It illustrates six consistent Ce 3d binding energy (BE) peaks for the rods with the former report on Ce^{4+}, signifying +4 was the main valence of rods in cerium [41]. The surface termination of oxide-based nanoparticles can be easily determined by the transmission electron microscopy (TEM) [130, 131]. The specific surfaces of catalytic CeO_2 nanostructures during a reversible beam induced redox reaction examined by combination of direct aberration corrected TEM and computational exit wavefunction restoration at ambient temperature.

Mesoporous ceria being versatile attracted researchers as catalysts and catalyst-support which possess increased dispersion of active secondary components and offer high surface area (**Figure 10**). Additionally, an issue which is mandatory to discuss is that as surfactant is removed during synthesis procedure, it shows its poor thermal stability at elevated temperatures precepted to be caused by collapsing of

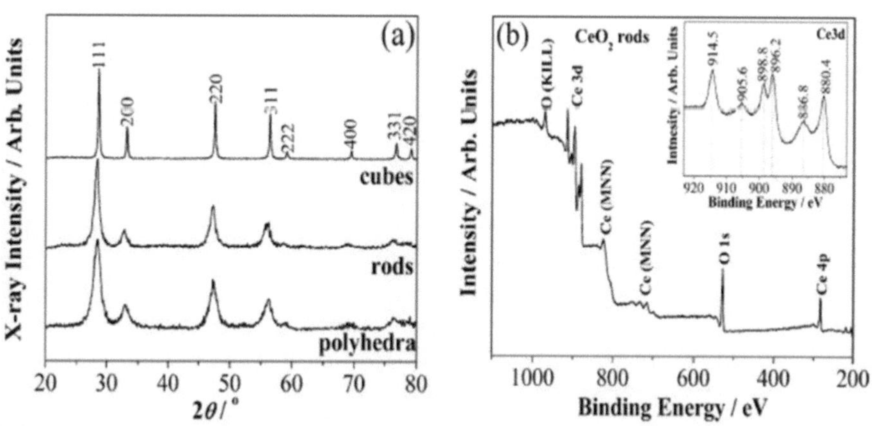

Figure 9.
(a) XRD patterns of CeO_2 nanorods, nanocubes and nanopolyhedra and (b) XPS wide spectrum of the CeO_2 nanorods. Reprinted with permission from Ref. [41]. Copyright 2005 American Chemical Society.

Figure 10.
Illustrative SEM images (a and b) and TEM images (c and d) of the flowerlike CeO_2 microspheres. Reprinted with permission from Ref. [72]. Copyright 2006 American Chemical Society.

structure. So, mesoporous CeO_2 has been developed by a novel hydrothermal method for high performance catalysts with excellent thermal stability [72, 132–135]. Hojo et al. analyze cationic reconstruction by using in-situ phase contrast HR-TEM with spherical aberration correction [136]. STEM, EELS, and theoretical calculations were performed to inspect the atomic structure of grain boundary which is selected as a model grain boundary in thin films of CeO_2 [38, 137, 138].

2.4 Cerium oxide catalyzed organic transformations

Organic synthesis occupied one of the most protruding places in the field of chemistry research. Additionally, the space of organocatalytic reactions is well-reviewed and widely examined. Further, initiating with certain reports to describe the organocatalytic applications of ceria nanostructures. Investigations on CeO_2 as catalytic support or recently as catalyst for conversions in organic reactions are liberated in accumulated manner. Although, the versatile behavior of this material is selectively observed in various catalytic applications. By Mars-van Krevelen mechanism, computer stimulation techniques were performed to predict the higher reactivity of ceria {110} and {100} surfaces towards carbon monoxide oxidation, which stated that CO first interacts with surface ceria oxygen and produce CO_2 by leaving an oxygen vacancy which is then filled with gas phase oxygen [139, 140]. The CO oxidation frequency turnover is higher on {110} as compare to {100} and {111} surfaces, presented as the opposite order of oxygen vacancy formation energy. CeO_2–ZnO composite catalyst utilized in hydrogen transfer reaction by Mishra et al. for cyclohexanone with isopropanol and it showed 51.3 mol% conversion of cyclohexanone [82]. Acetalization of cyclohexanone with methanol also reported by Rose et al. using different transition metals [141]. Tamizhdurai synthesized CeO_2 and inspect it with various spectroscopic and analytical techniques. Afterwards, its oxidation effect was investigated on benzyl alcohol which reveals better conversion and selectivity. The catalytic oxidation properties of ceria closely tied with its redox and oxygen storage behavior, and CO oxidation can work as a model reaction to probe the redox properties of CeO_2 [142] (**Scheme 1**).

Zhou et al. [143] explored oxidation of carbon monoxide over ceria nanostructures in their study where they compared nanorods and irregular nanoparticles of same surface area; the former attributed exposed planes of {100} and {110} surfaces with higher proportion has higher activity. This study initiated the investigation of CO oxidation with nanoparticles, and several investigations were followed which clearly establish the correlation between ceria shapes and CO oxidation as shown in **Figure 11** [143–150]. CeO_2 nanospheres fabricated sonochemically in 1-butyl-3-methylimidazolium bis(trifluoromethylsulfonyl)imide [C4mim] [Tf2N] show the best presentation for low-temperature CO oxidation [151].

CeO_2 nanoparticles explored by Deori K. for the para-xylene oxidation to terephthalic acid as a heterogenous catalyst (**Scheme 2**). The synthesis procedure is environmentally friendly, and water was used as a solvent during catalysis reaction. The ceria nanostructures which were synthesized acquire 15 nm sized particles and high surface area of 268 $m^2\ g^{-1}$ [152].

The advancement in catalytic performance of the cube shaped CeO_2 nanoparticles displayed by the conversion of benzyl alcohol (BA) and para-chlorobenzyl alcohol (PCBA) to their respective aldehydes (>99%) (**Schemes 3**). Accompanying, in toluene ($PhCH_3$) oxidation, this CeO_2 nanocube catalyst was found to be very effective, as well as being more effective than the nanorods. Besides, reusable property of CeO_2 nanocatalyst also proposed for several cycles which display obtainment of the desired products without any deterioration in selectivity and activity in all cases [153].

CeO_2 is actively used in hydrogenation reactions as a promoter or carrier of noble metal nanomaterials for many years [154, 155]. As compared to oxidation reaction on CeO_2, hydrogenation owing to the specific role of adjacent oxygen on

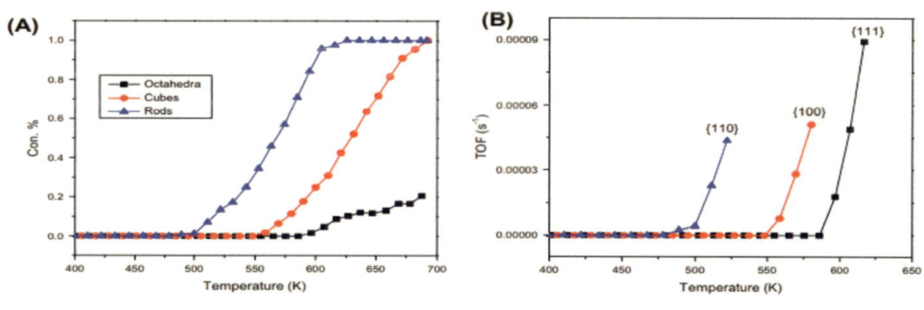

Figure 11.
(A) CO oxidation over ceria rods, cubes, and octahedra. (B) Comparison of CO turns over frequency oversurface planes (110), (100) and (111). Reprinted with permission from Ref. [143]. Copyright 2011 Elsevier.

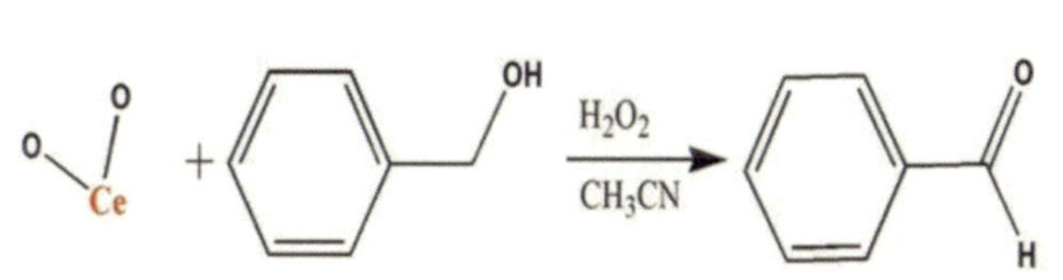

Scheme 1.
Selective oxidation of benzyl alcohol to benzaldehyde. Reprinted with permission from Ref. [142]. Copyright 2017 Scientific Reports.

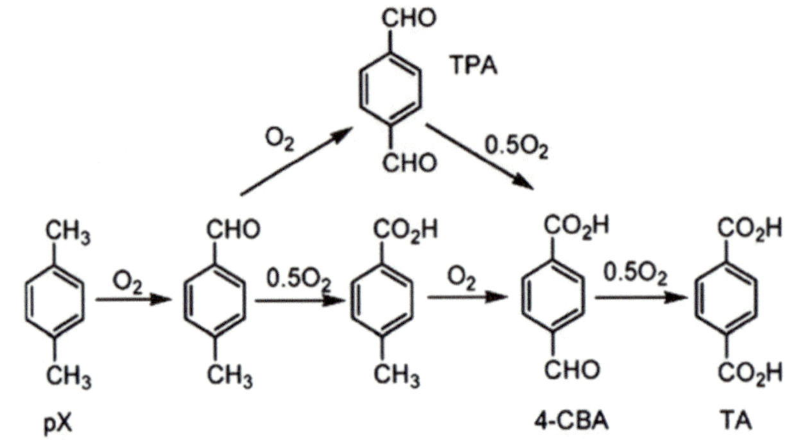

Scheme 2.
Oxidation process for para-xylene to terephthalic acid. Reprinted with permission from Ref. [152]. Copyright 2017 Royal Society of Chemistry.

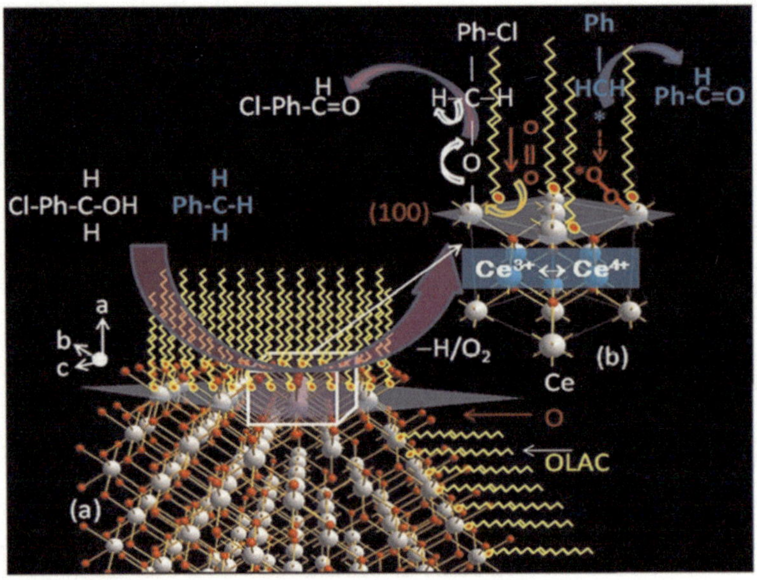

Scheme 3.
Representation of oxidation process of para-chlorobenzyl alcohol/toluene to benzaldehyde by ceria nanocubes. Reprinted with permission from Ref. [153]. Copyright 2017 Royal Society of Chemistry.

stabilizing hydroxyl intermediates is favored over low-vacancy surfaces whereas reverse effect applied for hydrogenation reactions, where nanoparticles are more active than nanocubes [156, 157]. Hydrogen activation on CeO_2 is often regarded as the limiting step of the reaction, even for other functional groups which includes substituted nitroarenes [28, 158]. As already discussed, partial hydrogenation over CeO_2, they have also been employed for the hydrogenation of olefins and carbonyl bonds. For example, a good yield of 1-butene can be obtained by reduction of 1,3-butadiene by Pd/CeO_2 catalyst supported on alumina [159]. Ceria nanostructures evidencing their value in other more complex organic reactions, apart from oxidations and hydrogenations, in advanced and controlled fabrication, promoting it as attractive and versatile nanocatalyst. Coupling reactions including aldol condensation (**Scheme 4**) [160, 161], Mannich reaction (**Scheme 5**) [162], Suzuki-Miyaura [163], Knoevenagel condensation (**Scheme 6**) [164] or Sonogashira cross couplings [165] have also been reported.

Yadav et al. [166, 167] have cast-off CeO_2 in synthesis. Recent Literature exposed that very few reagents have been reported for the bis-Michael addition reactions and most of the reagents, yielded the mono-Michael addition product [168]. Javad Safaei-Ghomi progressively synthesize CeO_2 nanoparticles and further utilize them by pseudo five-component reaction of acetylenedicarboxylates, phenylhydrazine and aromatic aldehydes in preparation of C-tethered bispyrazol-5-ols at 70°C in water [169]. In organic conversions, as carbon-carbon (C–C) bond formation reactions, the catalytic activity of free-CeO_2 NPs has not been studied extensively to the unsurpassed of our knowledge.

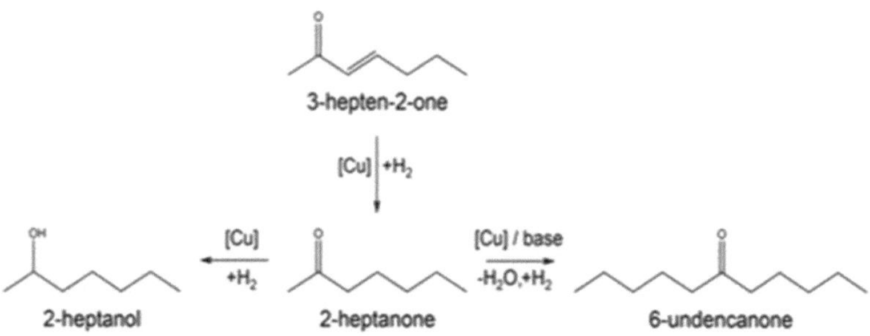

Scheme 4.
Cu supported catalysts on ceria-zirconia catalyzing cross-aldol condensation of acetone and n-butanol into aliphatic ketones. Reprinted with permission from Ref. [160]. Copyright 2017 Catalysts.

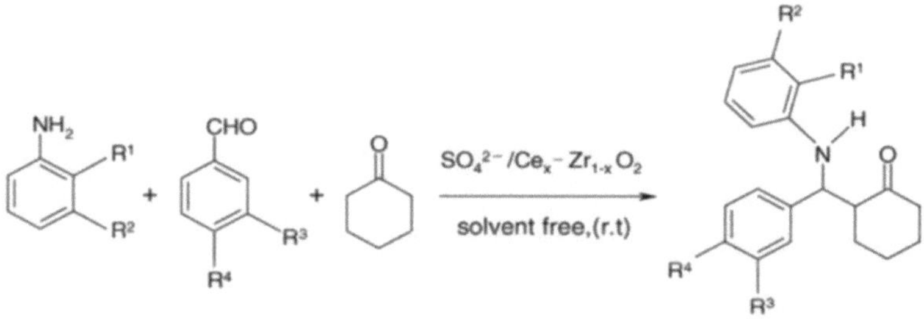

Scheme 5.
Mannich reaction catalyzed by Sulfated $Ce_xZr_{1-x}O_2$ catalyst. Reprinted with permission from Ref. [162]. Copyright 2006 Elsevier.

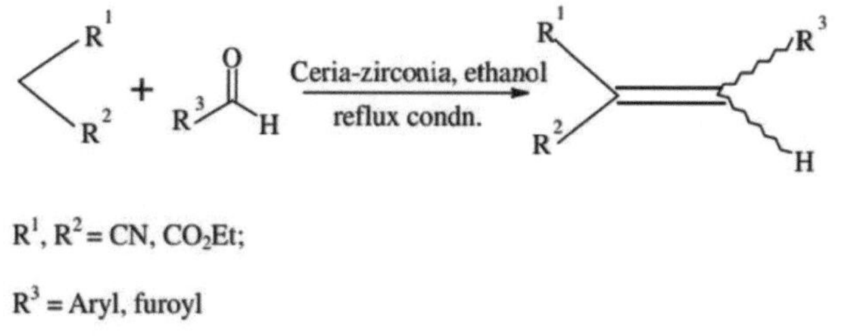

Scheme 6.
Possible mechanism of Knoevenagel condensation. Reprinted with permission from Ref. [164]. Copyright 2009 Elsevier.

3. Conclusion

Various synthetic strategies of Cerium oxide nanoparticles and their progress in the field of catalysis of organic transformations are selectively highlighted in this comprehensive chapter. We presented that at nano level, ceria structures are manipulated with different techniques which allows direct control over catalytic behavior in various reactions. Cerium oxide occupies widespread attention in research on new catalysts with improved properties for organic synthesis due to its very rich chemistry.

Ceria, firstly used by Ford Motor Company as an oxygen storage component, further stepping towards growth in its applications, as consider an "inert" support can stabilize metal nanoparticles which are actively practiced for its catalytic activities, that directly takes part in the reaction with lattice oxygen, afterwards a cocatalyst, and more recently a catalyst. Applications of Ce and Ce-based nanoparticles in different forms of catalysis with recent advances in their preparation methods are properly introduced in the chapter. The synthesis section included different preparation procedures such as hydrothermal, reverse micelle, Co-precipitation and sol-gel method for synthesis of Ce and Ce-based NPs and their characterization. These procedures show their importance in designing and development of Ce-based nanostructures by controlling the morphology of these nanosystems with featured catalytic applications ranging from organic transformations to photocatalysis, and so on. The catalytic improvements of ceria-based nanostructures followed two major directions. First, the surface area is increased with the enhancement of its thermal stability. Second, the nanostructures with well controlled shape and size are obtained by the advent of nanotechnology. While ceria-based materials effectively promoted several oxidation reactions as well as other emerging applications are also proposed.

In addition to this enormous applications of Ce nanocatalysts are reported for organic conversions such as hydrogenation, reduction, alkyne-azide cycloaddition, coupling reactions including A3, coupling. CeO_2 has a good feature as follows: their redox ability and the acid base properties whether they are doped with transition metals or alone. The activation of complex organic molecules with further possible transformation can possibly proceed due to these parameters. Certain acid-base and redox properties can adjust with various cerium-based mixed oxides and to control the number of active sites and their strength for the specific reaction. Latest advances in ceria nanocrystals synthesis with controlled morphologies such as nanocubes, nanorods, polyhedras, etc. should be leading towards encounter of novel catalysts with better selectivities and higher activities in catalysis and organic chemistry.

Acknowledgements

TA thanks to CSIR (Grant No. 01(2897)/17/EMR-II) and SERB-DST (Grant No. EMR/2016/001668), New Delhi, Govt. of India for financial support to research projects. FN and UF are thankful to UGC, New Delhi for Non-NET Research Fellowship.

Conflict of interest

The authors have no conflict of interest.

Author details

Farha Naaz, Umar Farooq and Tokeer Ahmad*
Department of Chemistry, Jamia Millia Islamia, New Delhi, India

*Address all correspondence to: tahmad3@jmi.ac.in

References

[1] Cole-Hamilton DJ. Homogeneous catalysis—New approaches to catalyst separation, recovery, and recycling. Science. 2003;**299**(5613):1702-1706. DOI: 10.1126/science.1081881

[2] Somorjai GA, Li Y. Selective nanocatalysis of organic transformation by metals: Concepts, model systems, and instruments. Topics in Catalysis. 2010;**53**(13–14):832-847. DOI: 10.1007/s11244-010-9511-y

[3] Thomas JM. Colloidal metals: Past, present and future. Pure and Applied Chemistry. 1988;**60**(10):1517-1528. DOI: 10.1351/pac198860101517

[4] Philippot K, Serp P. Concepts in nanocatalysis. In: Nanomater. Catal. 1st ed. New York: John-Wiley & Sons, Inc.; 2012. pp. 1-54. DOI: 10.1002/978352765 6875.ch1

[5] Freund PL, Spiro M. Colloidal catalysis: The effect of sol size and concentration. The Journal of Physical Chemistry. 1985;**89**(7):1074-1077. DOI: 10.1021/j100253a007

[6] Babu SG, Karvembu R. Copper based nanoparticles-catalyzed organic transformations. Catalysis Surveys from Asia. 2013;**17**(3–4):156-176. DOI: 10.1007/s10563-013-9159-2

[7] Gladysz JA. Recoverable catalysts. Ultimate goals, criteria of evaluation, and the green chemistry interface. Pure and Applied Chemistry. 2001;**73**(8): 1319-1324. DOI: 10.1351/pac200173081319

[8] Gladysz JA. Introduction: Recoverable catalysts and reagents perspective and prospective. Chemical Reviews. 2002;**102**(10):3215-3216. DOI: 10.1021/cr020068s

[9] Suramwar NV, Thakare SR, Khaty NT. Application of Nanomaterials as a Catalyst in Organic Synthesis. International Journal of Knowledge Engineering. 2012;**3**(1):98-99. DOI: 10.9735/0976-5816. ISSN 0976-5816

[10] Ahrens TJ. Global Earth Physic: A Handbook of Physical Constants. Washington, DC: American Geophysical Union; 1995. DOI: 10.14510/araproc.v0i0.1271

[11] Rao H, Jin Y, Fu H, Jiang Y, Zhao Y. A versatile and efficient ligand for copper-catalyzed formation of C– N, C–O, and P–C bonds: Pyrrolidine-2-phosphonic acid phenyl monoester. Chemistry–A European Journal. 2006; **12**(13):3636-3646. DOI: 10.1002/chem.200501473

[12] Kwong FY, Klapars A, Buchwald SL. Copper-catalyzed coupling of alkylamines and aryl iodides: An efficient system even in an air atmosphere. Organic Letters. 2002;**4**(4): 581-584. DOI: 10.1021/ol0171867

[13] Kwong FY, Buchwald SL. Mild and efficient copper-catalyzed amination of aryl bromides with primary alkylamines. Organic Letters. 2003;**5**(6): 793-796. DOI: 10.1021/ol0273396

[14] Wu TS, Zhou Y, Sabirianov RF, Mei WN, Soo YL, Cheung CL. X-ray absorption study of ceria nanorods promoting the disproportionation of hydrogen peroxide. Chemical Communications. 2016;**52**(28): 5003-5006. DOI: 10.1039/C5CC10643E

[15] Celardo I, Pedersen JZ, Traversa E, Ghibelli L. Pharmacological potential of cerium oxide nanoparticles. Nanoscale. 2011;**3**(4):1411-1420. DOI: 10.1039/C0NR00875C

[16] Esch F, Fabris S, Zhou L, Montini T, Africh C, Fornasiero P, et al. Electron localization determines defect

formation on ceria substrates. Science. 2005;**309**(5735):752-755. DOI: 10.1126/science.1111568

[17] Lawrence NJ, Brewer JR, Wang L, Wu TS, Wells-Kingsbury J, Ihrig MM, et al. Defect engineering in cubic cerium oxide nanostructures for catalytic oxidation. Nano Letters. 2011;**11**(7):2666-2671. DOI: 10.1021/nl200722z

[18] Liu W, Wang W, Tang K, Guo J, Ren Y, Wang S, et al. The promoting influence of nickel species in the controllable synthesis and catalytic properties of nickel–ceria catalysts. Catalysis Science & Technology. 2016;**6**(7):2427-2434. DOI: 10.1039/C5CY01241D

[19] Lu G, Linsebigler A, Yates JT Jr. Photooxidation of CH_3Cl on TiO_2 (110): A mechanism not involving H_2O. The Journal of Physical Chemistry. 1995;**99**(19):7626-7631. DOI: 10.1021/j100019a049

[20] Bai S, Shi B, Deng W, Dai Q, Wang X. Catalytic oxidation of 1, 2-dichloroethane over Al_2O_3–CeO_2 catalysts: Combined effects of acid and redox properties. RSC Advances. 2015;**5**(60):48916-48927. DOI: 10.1039/C5RA07405C

[21] Yuan Q, Duan HH, Li LL, Sun LD, Zhang YW, Yan CH. Controlled synthesis and assembly of ceria-based nanomaterials. Journal of Colloid and Interface Science. 2009;**335**(2):151-167. DOI: 10.1016/j.jcis.2009.04.007

[22] Huang W, Gao Y. Morphology-dependent surface chemistry and catalysis of CeO_2 nanocrystals. Catalysis Science & Technology. 2014;**4**(11):3772-3784. DOI: 10.1039/C4CY00679H

[23] Montini T, Melchionna M, Monai M, Fornasiero P. Fundamentals and catalytic applications of CeO_2-based materials. Chemical Reviews. 2016;**116**(10):5987-6041. DOI: 10.1021/acs.chemrev.5b00603

[24] Paier J, Penschke C, Sauer J. Oxygen defects and surface chemistry of ceria: Quantum chemical studies compared to experiment. Chemical Reviews. 2013;**113**(6):3949-3985. DOI: 10.1021/cr3004949

[25] Fan L, Wang C, Chen M, Zhu B. Recent development of ceria-based (nano) composite materials for low temperature ceramic fuel cells and electrolyte-free fuel cells. Journal of Power Sources. 2013;**234**:154-174. DOI: 10.1016/j.jpowsour.2013.01.138

[26] Sun C, Li H, Chen L. Nanostructured ceria-based materials: Synthesis, properties, and applications. Energy & Environmental Science. 2012;**5**(9):8475-8505. DOI: 10.1039/C2EE22310D

[27] Mogensen M, Sammes NM, Tompsett GA. Physical, chemical and electrochemical properties of pure and doped ceria. Solid State Ionics. 2000;**129**(1–4):63-94. DOI: 10.1016/S0167-2738(99)00318-5

[28] Vivier L, Duprez D. Ceria-based solid catalysts for organic chemistry. ChemSusChem. 2010;**3**(6):654-678. DOI: 10.1002/cssc.201000054

[29] Zhang D, Du X, Shi L, Gao R. Shape-controlled synthesis and catalytic application of ceria nanomaterials. Dalton Transactions. 2012;**41**(48):14455-14475. DOI: 10.1039/C2DT31759A

[30] Trovarelli A. Catalytic properties of ceria and CeO_2-containing materials. Catalysis Reviews. 1996;**38**(4):439-520. DOI: 10.1080/01614949608006464

[31] Mandoli C, Pagliari F, Pagliari S, Forte G, Di Nardo P, Licoccia S, et al. Stem cell aligned growth induced by CeO_2 nanoparticles in PLGA scaffolds

with improved bioactivity for regenerative medicine. Advanced Functional Materials. 2010;**20**(10): 1617-1624. DOI: 10.1002/adfm.200902363

[32] Xu C, Lin Y, Wang J, Wu L, Wei W, Ren J, et al. Nanoceria-triggered synergetic drug release based on CeO_2-capped mesoporous silica host–guest interactions and switchable enzymatic activity and cellular effects of CeO_2. Advanced Healthcare Materials. 2013; **2**(12):1591-1599. DOI: 10.1002/adhm.201200464

[33] Kim CK, Kim T, Choi IY, Soh M, Kim D, Kim YJ, et al. Ceria nanoparticles that can protect against ischemic stroke. Angewandte Chemie International Edition. 2012;**51**(44): 11039-11043. DOI: 10.1002/anie.201203780

[34] Furler P, Scheffe JR, Steinfeld A. Syngas production by simultaneous splitting of H_2O and CO_2 via ceria redox reactions in a high-temperature solar reactor. Energy & Environmental Science. 2012;**5**(3):6098-6103. DOI: 10.1039/C1EE02620H

[35] Scheffe JR, Welte M, Steinfeld A. Thermal reduction of ceria within an aerosol reactor for H_2O and CO_2 splitting. Industrial & Engineering Chemistry Research. 2014;**53**(6): 2175-2182. DOI: 10.1021/ie402620k

[36] Vyas S. Simulation of ceria: Bulk and surface defects [doctoral dissertation]. University of London; 1997

[37] Fronzi M, Soon A, Delley B, Traversa E, Stampfl C. Stability and morphology of cerium oxide surfaces in an oxidizing environment: A first-principles investigation. The Journal of Chemical Physics. 2009;**131**(10):104701. DOI: 10.1063/1.3191784

[38] Sayle TXT, Parker SC, Catlow CRA. The role of oxygen vacancies on ceria surfaces in the oxidation of carbon monoxide. Surface Science. 1994; **316**(3):329-336. DOI: 10.1016/0039-6028(94)91225-4

[39] Kaneko K, Inoke K, Freitag B, Hungria AB, Midgley PA, Hansen TW, et al. Structural and morphological characterization of cerium oxide nanocrystals prepared by hydrothermal synthesis. Nano Letters. 2007;**7**(2): 421-425. DOI: 10.1021/nl062677b

[40] Otsuka-Yao-Matsuo S, Omata T, Izu N, Kishimoto H. Oxygen release behavior of $CeZrO_4$ powders and appearance of new compounds k and t. Journal of Solid State Chemistry. 1998; **138**(1):47-54. DOI: 10.1006/jssc.1998.7753

[41] Mai HX, Sun LD, Zhang YW, Si R, Feng W, Zhang HP, et al. Shape-selective synthesis and oxygen storage behavior of ceria nanopolyhedra, nanorods, and nanocubes. The Journal of Physical Chemistry B. 2005;**109**(51): 24380-24385. DOI: 10.1021/jp055584b

[42] Trovarelli A, Llorca J. Ceria catalysts at nanoscale: How do crystal shapes shape catalysis? ACS Catalysis. 2017; **7**(7):4716-4735. DOI: 10.1021/acscatal.7b01246

[43] Singh SB, Tandon PK. Catalysis: A brief review on nano-catalyst. Journal of Energy and Chemical Engineering. 2014;**2**(3):106-115. DOI: 10.4172/2157-7439-C1-058

[44] Chen L, Fleming P, Morris V, Holmes JD, Morris MA. Size-related lattice parameter changes and surface defects in ceria nanocrystals. The Journal of Physical Chemistry C. 2010; **114**(30):12909-12919. DOI: 10.1021/jp1031465

[45] Hua G, Zhang L, Fei G, Fang M. Enhanced catalytic activity induced by defects in mesoporous ceria nanotubes. Journal of Materials Chemistry. 2012;

22(14):6851-6855. DOI: DOI :10.1039/c2jm13610d

[46] Tang CC, Bando Y, Liu BD, Golberg D. Cerium oxide nanotubes prepared from cerium hydroxide nanotubes. Advanced Materials. 2005;**17**(24): 3005-3009. DOI: 10.1002/adma.200501557

[47] Averill BA. Principles of General Chemistry2012. pp. 1403-1521

[48] Stratton TG, Tuller HL. Thermodynamic and transport studies of mixed oxides. The CeO_2–UO_2 system. Journal of the Chemical Society, Faraday Transactions 2: Molecular and Chemical Physics. 1987;**83**(7):1143-1156. DOI: 10.1039/F29878301143

[49] Tuller HL, Bishop SR. Tailoring material properties through defect engineering. Chemistry Letters. 2010; **39**(12):1226-1231. DOI: 10.1246/cl.2010.1226

[50] Li J, Zhang Z, Tian Z, Zhou X, Zheng Z, Ma Y, et al. Low pressure induced porous nanorods of ceria with high reducibility and large oxygen storage capacity: Synthesis and catalytic applications. Journal of Materials Chemistry A. 2014;**2**(39):16459-16466. DOI: 10.1039/C4TA03718A

[51] Bernal S, Calvino JJ, Cauqui MA, Gatica JM, Larese C, Omil JP, et al. Some recent results on metal/support interaction effects in NM/CeO_2 (NM: noble metal) catalysts. Catalysis Today. 1999;**50**(2):175-206. DOI: 10.1016/S0920-5861(98)00503-3

[52] Nilsson A, Pettersson LGM. Chemical bonding on surfaces probed by X-ray emission spectroscopy and density functional theory. Surface Science Reports. 2004;**55**(2–5):49-167. DOI: 10.1016/j.surfrep.2004.06.002

[53] Lawrence NJ. Synthesis and catalytic activity of nanostructured cerium oxide [dissertation]. 2010

[54] Aneggi E, Boaro M, de Leitenburg C, Dolcetti G, Trovarelli A. Insights into the redox properties of ceria-based oxides and their implications in catalysis. Journal of Alloys and Compounds. 2006;**408**:1096-1102. DOI: 10.1016/j.jallcom.2004.12.113

[55] Wang G, Wang L, Fei X, Zhou Y, Sabirianov RF, Mei WN, et al. Probing the bifunctional catalytic activity of ceria nanorods towards the cyanosilylation reaction. Catalysis Science & Technology. 2013;**3**(10): 2602-2609. DOI: 10.1039/C3CY00196B

[56] Khaladji J, Peltier M, Rhone-Poulenc Industries. Rare earth polishing compositions. U.S. Patent 4,942,697; 1990

[57] Liu L, Yao Z, Deng Y, Gao F, Liu B, Dong L. Morphology and crystal-plane effects of nanoscale ceria on the activity of CuO/CeO_2 for NO reduction by CO. ChemCatChem. 2011;**3**(6):978-989. DOI: 10.1002/cctc.201000320

[58] Sabot J-L, Maestro P. Kirk-Othmer Encyclopedia of Chemical Technology. New York: John Wiley & Sons, Inc.; 2000. DOI: 10.1007/978-90-481-8679-2

[59] Schermanz K, Sagar A, TreibacherIndustrie AG. Ceria zirconia alumina composition with enhanced thermal stability. U.S. Patent 9,475,035; 2016

[60] Wang JF, Rohm and Haas Electronic Materials CMP Inc. Oxide particles and method for producing them. U.S. Patent 5,389,352; 1995

[61] Hanawa K, Mochizuki N, Ueda N, Mitsui Mining and Smelting Co Ltd. Cerium oxide ultrafine particles and method for preparing the same. U.S. Patent 5,938,837; 1999

[62] Rogal J, Reuter K, Scheffler M. CO oxidation at Pd (100): A first-principles constrained thermodynamics study.

Physical Review B. 2007;**75**(20):205433. DOI: 10.1103/PhysRevB.75.205433

[63] Mavrikakis M, Hammer B, Nørskov JK. Effect of strain on the reactivity of metal surfaces. Physical Review Letters. 1998;**81**(13):2819. DOI: DOI : 10.1103/ PhysRevLett.81.2819

[64] Grabow L, Xu Y, Mavrikakis M. Lattice strain effects on CO oxidation on Pt (111). Physical Chemistry Chemical Physics. 2006;**8**(29):3369-3374. DOI: 10.1039/B606131A

[65] Greeley J, Krekelberg WP, Mavrikakis M. Strain-induced formation of subsurface species in transition metals. Angewandte Chemie. 2004; **116**(33):4396-4400. DOI: 10.1002/ ange.200454062

[66] Venezia AM, Pantaleo G, Longo A, Di Carlo G, Casaletto MP, Liotta FL, et al. Relationship between structure and CO oxidation activity of ceria-supported gold catalysts. The Journal of Physical Chemistry B. 2005;**109**(7): 2821-2827. DOI: 10.1021/jp045928i

[67] Senanayake SD, Stacchiola D, Rodriguez JA. Unique properties of ceria nanoparticles supported on metals: Novel inverse ceria/copper catalysts for CO oxidation and the water-gas shift reaction. Accounts of Chemical Research. 2013;**46**(8):1702-1711. DOI: 10.1021/ar300231p

[68] Wang F. Structural and electronic properties of noble metals on metal oxide surfaces [doctoral dissertation]. 2011

[69] White RJ, Luque R, Budarin VL, Clark JH, Macquarrie DJ. Supported metal nanoparticles on porous materials. Methods and applications. Chemical Society Reviews. 2009;**38**(2):481-494. DOI: 10.1039/B802654H

[70] Vantomme A, Yuan ZY, Du G, Su BL. Surfactant-assisted large-scale preparation of crystalline CeO_2 nanorods. Langmuir. 2005;**21**(3): 1132-1135. DOI: 10.1021/la047751p

[71] Lu X, Zhai T, Cui H, Shi J, Xie S, Huang Y, et al. Redox cycles promoting photocatalytic hydrogen evolution of CeO_2 nanorods. Journal of Materials Chemistry. 2011;**21**(15):5569-5572. DOI: 10.1039/C0JM04466K

[72] Sun C, Sun J, Xiao G, Zhang H, Qiu X, Li H, et al. Mesoscale organization of nearly monodisperse flowerlike ceria microspheres. The Journal of Physical Chemistry B. 2006;**110**(27): 13445-13452. DOI: 10.1021/jp062179r

[73] Xie X, Li Y, Liu ZQ, Haruta M, Shen W. Low-temperature oxidation of CO catalysed by Co_3O_4 nanorods. Nature. 2009;**458**(7239):746. DOI: 10.1038/ nature07877

[74] Yang HG, Sun CH, Qiao SZ, Zou J, Liu G, Smith SC, et al. Anatase TiO_2 single crystals with a large percentage of reactive facets. Nature. 2008;**453**(7195): 638. DOI: 10.1038/nature06964

[75] Leng M, Liu M, Zhang Y, Wang Z, Yu C, Yang X, et al. Polyhedral 50-facet Cu_2O microcrystals partially enclosed by {311} high-index planes: Synthesis and enhanced catalytic CO oxidation activity. Journal of the American Chemical Society. 2010;**132**(48): 17084-17087. DOI: 10.1021/ja106788x

[76] Wang ZL, Feng X. Polyhedral shapes of CeO_2 nanoparticles. The Journal of Physical Chemistry B. 2003; **107**(49):13563-13566. DOI: 10.1021/ jp036815m

[77] Hirano M, Fukuda Y, Iwata H, Hotta Y, Inagaki M. Preparation and spherical agglomeration of crystalline cerium (IV) oxide nanoparticles by thermal hydrolysis. Journal of the American Ceramic Society. 2000;**83**(5):1287-1289. DOI: 10.1111/j.1151-2916. 2000.tb01371.x

[78] Zhou F, Zhao X, Xu H, Yuan C. CeO_2 spherical crystallites: Synthesis, formation mechanism, size control, and electrochemical property study. The Journal of Physical Chemistry C. 2007; **111**(4):1651-1657. DOI: 10.1021/jp0660435

[79] Xia B, Lenggoro IW, Okuyama K. Synthesis of CeO_2 nanoparticles by salt-assisted ultrasonic aerosol decomposition. Journal of Materials Chemistry. 2001;**11**(12):2925-2927. DOI: 10.1039/B105548H

[80] Wang S, Zhao L, Wang W, Zhao Y, Zhang G, Ma X, et al. Morphology control of ceria nanocrystals for catalytic conversion of CO_2 with methanol. Nanoscale. 2013;**5**(12): 5582-5588. DOI: 10.1039/C3NR00831B

[81] Dai Q, Bai S, Li H, Liu W, Wang X, Lu G. Template-free and non-hydrothermal synthesis of CeO_2 nanosheets via a facile aqueous-phase precipitation route with catalytic oxidation properties. CrystEngComm. 2014;**16**(42):9817-9827. DOI: 10.1039/C4CE01436G

[82] Zhou K, Wang X, Sun X, Peng Q, Li Y. Enhanced catalytic activity of ceria nanorods from well-defined reactive crystal planes. Journal of Catalysis. 2005;**229**(1):206-212. DOI: 10.1016/j.jcat.2004.11.004

[83] Dai Q, Huang H, Zhu Y, Deng W, Bai S, Wang X, et al. Catalysis oxidation of 1, 2-dichloroethane and ethyl acetate over ceria nanocrystals with well-defined crystal planes. Applied Catalysis B: Environmental. 2012;**117**:360-368. DOI: 10.1016/j.apcatb.2012.02.001

[84] Choi M, Na K, Kim J, Sakamoto Y, Terasaki O, Ryoo R. Stable single-unit-cell nanosheets of zeolite MFI as active and long-lived catalysts. Nature. 2009; **461**(7261):246. DOI: 10.1038/nature08288

[85] Kim JE, Carroll PJ, Schelter EJ. Bidentate nitroxide ligands stable toward oxidative redox cycling and their complexes with cerium and lanthanum. Chemical Communications. 2015; **51**(81):15047-15050. DOI: 10.1039/C5CC06052D

[86] Qi J, Chen J, Li G, Li S, Gao Y, Tang Z. Facile synthesis of core–shell Au@CeO_2 nanocomposites with remarkably enhanced catalytic activity for CO oxidation. Energy & Environmental Science. 2012;**5**(10):8937-8941. DOI: 10.1039/C2EE22600F

[87] Mitsudome T, Mikami Y, Matoba M, Mizugaki T, Jitsukawa K, Kaneda K. Design of a silver–cerium dioxide core–shell nanocomposite catalyst for chemoselective reduction reactions. Angewandte Chemie International Edition. 2012;**51**(1):136-139

[88] Saada R, Kellici S, Heil T, Morgan D, Saha B. Greener synthesis of dimethyl carbonate using a novel ceria–zirconia oxide/graphene nanocomposite catalyst. Applied Catalysis B: Environmental. 2015;**168**:353-362. DOI: 10.1016/j.apcatb.2014.12.013

[89] Cargnello M, Wieder NL, Montini T, Gorte RJ, Fornasiero P. Synthesis of dispersible Pd@CeO_2 core-shell nanostructures by self-assembly. Journal of the American Chemical Society. 2009;**132**(4):1402-1409. DOI: 10.1021/ja909131k

[90] Yamada Y, Tsung CK, Huang W, Huo Z, Habas SE, Soejima T, et al. Nanocrystal bilayer for tandem catalysis. Nature Chemistry. 2011;**3**(5): 372. DOI: 10.1038/nchem.1018

[91] Grinter DC, Ithnin R, Pang CL, Thornton G. Defect structure of ultrathin ceria films on Pt (111): Atomic views from scanning tunnelling microscopy. The Journal of Physical Chemistry C. 2010;**114**(40): 17036-17041. DOI: 10.1021/jp102895k

[92] Rodriguez JA, Grinter DC, Liu Z, Palomino RM, Senanayake SD. Ceria-based model catalysts: Fundamental studies on the importance of the metal–ceria interface in CO oxidation, the water-gas shift, CO_2 hydrogenation, and methane and alcohol reforming. Chemical Society Reviews. 2017;**46**(7): 1824-1841. DOI: 10.1039/C6CS00863A

[93] Harding C, Habibpour V, Kunz S, Farnbacher ANS, Heiz U, Yoon B, et al. Control and manipulation of gold nanocatalysis: Effects of metal oxide support thickness and composition. Journal of the American Chemical Society. 2008;**131**(2):538-548. DOI: 10.1021/ja804893b

[94] Campbell CT, Parker SC, Starr DE. The effect of size-dependent nanoparticle energetics on catalyst sintering. Science. 2002;**298**(5594): 811-814. DOI: 10.1126/science.1075094

[95] Renaud G, Lazzari R, Revenant C, Barbier A, Noblet M, Ulrich O, et al. Real-time monitoring of growing nanoparticles. Science. 2003;**300**(5624): 1416-1419. DOI: 10.1126/science.1082146

[96] Conner WC Jr, Falconer JL. Spillover in heterogeneous catalysis. Chemical Reviews. 1995;**95**(3):759-788. DOI: 10.1021/cr00035a014

[97] Libuda J, Freund HJ. Molecular beam experiments on model catalysts. Surface Science Reports. 2005;**57**(7–8): 157-298. DOI: 10.1016/j.surfrep.2005.03.002

[98] Vayssilov GN, Lykhach Y, Migani A, Staudt T, Petrova GP, Tsud N, et al. Support nanostructure boosts oxygen transfer to catalytically active platinum nanoparticles. Nature Materials. 2011; **10**(4):310. DOI: 10.1038/nmat2976

[99] Bond GC, Thompson DT. Catalysis by gold. Catalysis Reviews. 1999;**41**(3–4): 319-388. DOI: 10.1081/CR-100101171

[100] Bunluesin T, Gorte RJ, Graham GW. CO oxidation for the characterization of reducibility in oxygen storage components of three-way automotive catalysts. Applied Catalysis B: Environmental. 1997;**14** (1–2):105-115. DOI: 10.1016/S0926-3373(97)00016-7

[101] Putna ES, Bunluesin T, Fan XL, Gorte RJ, Vohs JM, Lakis RE, et al. Ceria films on zirconia substrates: Models for understanding oxygen-storage properties. Catalysis Today. 1999;**50**(2): 343-352. DOI: 10.1016/S0920-5861(98)00514-8

[102] Cordatos H, Bunluesin T, Stubenrauch J, Vohs JM, Gorte RJ. Effect of ceria structure on oxygen migration for Rh/ceria catalysts. The Journal of Physical Chemistry. 1996; **100**(2):785-789. DOI: 10.1021/jp952050+

[103] Mamontov E, Egami T, Brezny R, Koranne M, Tyagi S. Lattice defects and oxygen storage capacity of nanocrystalline ceria and ceria-zirconia. The Journal of Physical Chemistry B. 2000;**104**(47):11110-11116. DOI: 10.1021/jp0023011

[104] Fornasiero P, Dimonte R, Rao GR, Kaspar J, Meriani S, Trovarelli AO, et al. Rh-loaded CeO_2-ZrO_2 solid-solutions as highly efficient oxygen exchangers: Dependence of the reduction behavior and the oxygen storage capacity on the structural-properties. Journal of Catalysis. 1995;**151**(1):168-177. DOI: 10.1006/jcat.1995.1019

[105] Fornasiero P, Balducci G, Di Monte R, Kašpar J, Sergo V, Gubitosa G, et al. Modification of the redox behaviour of CeO_2 induced by structural doping with ZrO_2. Journal of Catalysis. 1996; **164**(1):173-183. DOI: 10.1006/jcat.1996.0373

[106] Vidmar P, Fornasiero P, Kaspar J, Gubitosa G, Graziani M, Catal J. Effects of trivalent dopants on the redox

properties of $Ce_{0.6}Zr_{0.4}O_2$ mixed oxide. Journal of Catalysis. 1997;**171**(1): 160-168. DOI: 10.1006/jcat.1987.1784

[107] Fornasiero P, Kašpar J, Sergo V, Graziani M. Redox behavior of high-surface-area Rh-, Pt-, and Pd-loaded $Ce_{0.5}Zr_{0.5}O_2$ mixed oxide. Journal of Catalysis. 1999;**182**(1):56-69. DOI: 10.1006/jcat.1998.2321

[108] Vlaic G, Fornasiero P, Geremia S, Kašpar J, Graziani M. Relationship between the zirconia-promoted reduction in the Rh-loaded $Ce_{0.5}Zr_{0.5}O_2$ mixed oxide and the Zr–O local structure. Journal of Catalysis. 1997; **168**(2):386-392. DOI: 10.1006/jcat.1997.1644

[109] Vlaic G, Di Monte R, Fornasiero P, Fonda E, Kašpar J, Graziani M. Redox property–local structure relationships in the Rh-loaded CeO_2–ZrO_2 mixed oxides. Journal of Catalysis. 1999;**182**(2): 378-389. DOI: 10.1006/jcat.1998.2335

[110] Fornasiero P, Fonda E, Di Monte R, Vlaic G, Kašpar J, Graziani M. Relationships between structural/textural properties and redox behavior in $Ce_{0.6}Zr_{0.4}O_2$ mixed oxides. Journal of Catalysis. 1999;**187**(1):177-185. DOI: 10.1006/jcat.1999.2589

[111] Rodriguez JA, Liu P, Graciani J, Senanayake SD, Grinter DC, Stacchiola D, et al. Inverse oxide/metal catalysts in fundamental studies and practical applications: A perspective of recent developments. The Journal of Physical Chemistry Letters. 2016;**7**(13): 2627-2639. DOI: 10.1021/acs.jpclett.6b00499

[112] Rodriguez JA, Liu P, Graciani J, Senanayake SD, Grinter DC, Stacchiola D, et al. Inverse oxide/metal catalysts in fundamental studies and practical applications: A perspective of recent developments. The Journal of Physical Chemistry Letters. 2016;**7**(13):2627-2639. DOI: 10.1021/acs.jpclett.6b00499

[113] Raimondi F, Scherer GG, Kötz R, Wokaun A. Nanoparticles in energy technology: Examples from electrochemistry and catalysis. Angewandte Chemie International Edition. 2005;**44**(15):2190-2209. DOI: 10.1002/anie.200460466

[114] Vayssilov GN, Migani A, Neyman K. Density functional modeling of the interactions of platinum clusters with CeO_2 nanoparticles of different size. The Journal of Physical Chemistry C. 2011; **115**(32):16081-16086. DOI: 10.1021/jp204222k

[115] Wang JH, Liu M, Lin MC. Oxygen reduction reactions in the SOFC cathode of Ag/CeO_2. Solid State Ionics. 2006;**177** (9–10):939-947. DOI: 10.1016/j.ssi.2006.02.029

[116] Vári G, Óvári L, Kiss J, Kónya Z. LEIS and XPS investigation into the growth of cerium and cerium dioxide on Cu (111). Physical Chemistry Chemical Physics. 2015;**17**(7):5124-5132. DOI: 10.1039/C4CP05179C

[117] Zhou Y. Nanostructured cerium oxide-based catalysts: Synthesis, physical properties, and catalytic performance [doctoral dissertation]. 2015

[118] Eyring L. The binary lanthanide oxides: Synthesis and identification. In: Synthesis of Lanthanide and Actinide Compounds. Dordrecht: Springer; 1991. pp. 187-224. DOI: 10.1007/978-94-011-3758-4-8

[119] De Leitenburg C, Trovarelli A, Kašpar J. A temperature-programmed and transient kinetic study of CO_2 activation and methanation over CeO_2 supported noble metals. Journal of Catalysis. 1997;**166**(1):98-107. DOI: 10.1006/jcat.1997.1498

[120] Audebrand N, Auffrédic JP, Louër D. An X-ray powder diffraction study of

the microstructure and growth kinetics of nanoscale crystallites obtained from hydrated cerium oxides. Chemistry of Materials. 2000;**12**(6):1791-1799. DOI: 10.1021/cm001013e

[121] Nakane S, Tachi T, Yoshinaka M, Hirota K, Yamaguchi O. Characterization and sintering of reactive cerium (IV) oxide powders prepared by the hydrazine method. Journal of the American Ceramic Society. 1997;**80**(12):3221-3224. DOI: 10.1111/j.1151-2916.1997.tb03255.x

[122] Abimanyu H, Ahn BS, Kim CS, Yoo KS. Preparation and characterization of MgO-CeO_2 mixed oxide catalysts by modified coprecipitation using ionic liquids for dimethyl carbonate synthesis. Industrial & Engineering Chemistry Research. 2007;**46**(24):7936-7941. DOI: 10.1021/ie070528d

[123] Ward DA, Ko EI. Preparing catalytic materials by the sol-gel method. Industrial & Engineering Chemistry Research. 1995;**34**(2): 421-433. DOI: 10.1021/ie00041a001

[124] Gnanam S, Rajendran V. Synthesis of CeO_2 or α-Mn_2O_3 nanoparticles via sol–gel process and their optical properties. Journal of Sol-Gel Science and Technology. 2011;**58**(1):62-69. DOI: 10.1007/s10971-010-2356-9

[125] Masui T, Fujiwara K, Machida KI, Adachi GY, Sakata T, Mori H. Characterization of cerium (IV) oxide ultrafine particles prepared using reversed micelles. Chemistry of Materials. 1997;**9**(10):2197-2204. DOI: 10.1021/cm970359v

[126] Vaidya S, Agarwal S, Ahmad T, Ganguli AK. Nanocrystalline oxalate/carbonate precursors of Ce and Zr and their decompositions to CeO_2 and ZrO_2 nanoparticles. Journal of American Ceramic Society. 2007;**90**(3):863-869. DOI: 10.1111/j.1551-2916.2007.01484.x

[127] Ganguli AK, Vaidya S, Ahmad T. Synthesis of nanocrystalline materials through reverse micelles: A versatile methodology for synthesis of complex metal oxides. Bulletin of Materials Science. 2008;**31**(3):415-419. DOI: 10.1007/s12034-008-0065-6

[128] Ahmad T, Shahazad M, Ubaidullah M, Ahmed J, Khan A, El-Toni AM. Structural characterization and dielectric properties of ceria–titania nanocomposites in low ceria region. Materials Research Express. 2017;**4**(12):125016. DOI: 10.1088/2053-1591/aa9c51

[129] Ahmad T, Shahazad M, Ubaidullah M, Ahmed J. Synthesis, characterization and dielectric properties of TiO_2-CeO_2 ceramic nanocomposites at low titania concentration. Bulletin of Materials Science. 2018;**41**(4):99. DOI: 10.1007/s12034-018-1616-0

[130] Ahmed J, Ahmad T, Ramanujachary KV, Lofland SE, Ganguli AK. Development of a microemulsion-based process for synthesis of cobalt (Co) and cobalt oxide (Co_3O_4) nanoparticles from submicrometer rods of cobalt oxalate. Journal of Colloid and Interface Science. 2008;**321**(2):434-441. DOI: 10.1016/j.jcis.2008.01.052

[131] Ganguly A, Trinh P, Ramanujachary KV, Ahmad T, Mugweru A, Ganguli AK. Reverse micellar based synthesis of ultrafine MgO nanoparticles (8–10 nm): Characterization and catalytic properties. Journal of Colloid and Interface Science. 2011;**353**(1):137-142. DOI: 10.1016/j.jcis.2010.09.041

[132] Lyons DM, McGrath JP, Morris MA. Surface studies of ceria and mesoporous ceria powders by solid-state 1H MAS NMR. The Journal of Physical Chemistry B. 2003;**107**(19):4607-4617. DOI: 10.1021/jp0341570

[133] Terribile D, Trovarelli A, Llorca J, de Leitenburg C, Dolcetti G. The synthesis and characterization of mesoporous high-surface area ceria prepared using a hybrid organic/inorganic route. Journal of Catalysis. 1998;**178**(1):299-308. DOI: 10.1006/jcat.1998.2152

[134] Carreon MA, Guliants VV. Ordered meso-and macroporous binary and mixed metal oxides. European Journal of Inorganic Chemistry. 2005;**2005**(1): 27-43. DOI: 10.1002/ejic.200400675

[135] Hojo H, Mizoguchi T, Ohta H, Findlay SD, Shibata N, Yamamoto T, et al. Atomic structure of a CeO_2 grain boundary: The role of oxygen vacancies. Nano Letters. 2010;**10**(11):4668-4672. DOI: 10.1021/nl1029336

[136] Bhatta UM, Ross IM, Sayle TX, Sayle DC, Parker SC, Reid D, et al. Cationic surface reconstructions on cerium oxide nanocrystals: An aberration-corrected HRTEM study. ACS Nano. 2012;**6**(1):421-430. DOI: 10.1021/nn2037576

[137] Haruta M. Nanoparticulate gold catalysts for low-temperature CO oxidation. Journal of New Materials for Electrochemical Systems. 2004;**7**: 163-172. DOI: 10.1016/j.apcatb.2003.11.003

[138] Sun C, Li H, Chen L. Study of flowerlike CeO_2 microspheres used as catalyst supports for CO oxidation reaction. Journal of Physics and Chemistry of Solids. 2007;**68**(9): 1785-1790. DOI: 10.1016/j.jpcs.2007.05.005

[139] Conesa J. Computer modeling of surfaces and defects on cerium dioxide. Surface Science. 1995;**339**(3): 337-352. DOI: 10.1016/0039-6028(95)00595-1

[140] Piumetti M, Andana T, Bensaid S, Russo N, Fino D, Pirone R. Study on the CO oxidation over ceria-based nanocatalysts. Nanoscale Research Letters. 2016;**11**(1):165. DOI: 10.1186/s11671-016-1375-z

[141] Yan L, Yu R, Chen J, Xing X. Template-free hydrothermal synthesis of CeO_2 nano-octahedrons and nanorods: Investigation of the morphology evolution. Crystal Growth and Design. 2008;**8**(5):1474-1477. DOI: 10.1021/cg800117v

[142] Tamizhdurai P, Sakthinathan S, Chen SM, Shanthi K, Sivasanker S, Sangeetha P. Environmentally friendly synthesis of CeO_2 nanoparticles for the catalytic oxidation of benzyl alcohol to benzaldehyde and selective detection of nitrite. Scientific Reports. 2017;7:46372. DOI: 10.1038/srep46372

[143] Wu Z, Li M, Overbury SH. On the structure dependence of CO oxidation over CeO_2 nanocrystals with well-defined surface planes. Journal of Catalysis. 2012;**285**(1):61-73. DOI: 10.1016/j.jcat.2011.09.011

[144] Wang X, Jiang Z, Zheng B, Xie Z, Zheng L. Synthesis and shape-dependent catalytic properties of CeO_2 nanocubes and truncated octahedra. CrystEngComm. 2012;**14**(22):7579-7582. DOI: 10.1039/C2CE25333J

[145] Zhou HP, Zhang YW, Si R, Zhang LS, Song WG, Yan CH. Dimension-manipulated ceria nanostructures (0D uniform nanocrystals, 2D polycrystalline assembly, and 3D mesoporous framework) from cerium octylate precursor in solution phases and their CO oxidation activities. The Journal of Physical Chemistry C. 2008; **112**(51):20366-20374. DOI: 10.1021/jp807091n

[146] Li J, Lu G, Li H, Wang Y, Guo Y, Guo Y. Facile synthesis of 3D flowerlike CeO_2 microspheres under mild condition with high catalytic performance for CO oxidation. Journal

of Colloid and Interface Science. 2011; **360**(1):93-99. DOI: 10.1016/j.jcis.2011.04.052

[147] Cui R, Lu W, Zhang L, Yue B, Shen S. Template-free synthesis and self-assembly of CeO_2 nanospheres fabricated with foursquare nanoflakes. The Journal of Physical Chemistry C. 2009;**113**(52):21520-21525. DOI: 10.1021/jp9065168

[148] Renuka NK, Praveen AK, Aniz CU. Ceria rhombic microplates: Synthesis, characterization and catalytic activity. Microporous and Mesoporous Materials. 2013;**169**:35-41. DOI: 10.1016/j.micromeso.2012.10.010

[149] Li J, Zhang Z, Gao W, Zhang S, Ma Y, Qu Y. Pressure regulations on the surface properties of CeO_2 nanorods and their catalytic activity for CO oxidation and nitrile hydrolysis reactions. ACS Applied Materials & Interfaces. 2016; **8**(35):22988-22996. DOI: 10.1021/acsami.6b05343

[150] Lykaki M, Pachatouridou E, Iliopoulou E, Carabineiro SA, Konsolakis M. Impact of the synthesis parameters on the solid-state properties and the CO oxidation performance of ceria nanoparticles. RSC Advances. 2017;7(10):6160-6169. DOI: 10.1039/C6RA26712B

[151] Pan C, Zhang D, Shi L. CTAB assisted hydrothermal synthesis, controlled conversion and CO oxidation properties of CeO_2 nanoplates, nanotubes, and nanorods. Journal of Solid State Chemistry. 2008;**181**(6): 1298-1306. DOI: 10.1016/j.jssc.2008.02.011

[152] Deori K, Gupta D, Saha B, Awasthi SK, Deka S. Introducing nanocrystalline CeO_2 as heterogeneous environmentally friendly catalyst for the aerobic oxidation of para-xylene to terephthalic acid in water. Journal of Materials Chemistry A. 2013;**1**(24):7091-7099. DOI: 10.1039/C3TA01590D

[153] Deori K, Kalita C, Deka S. (100) surface-exposed CeO_2 nanocubes as an efficient heterogeneous catalyst in the tandem oxidation of benzyl alcohol, para-chlorobenzyl alcohol and toluene to the corresponding aldehydes selectively. Journal of Materials Chemistry A. 2015;**3**(13):6909-6920. DOI: 10.1039/C4TA06547F

[154] Segura Y, Lopez N, Perez-Ramirez J. Origin of the superior hydrogenation selectivity of gold nanoparticles in alkyne + alkene mixtures: Triple-versus double-bond activation. Journal of Catalysis. 2007;**247**(2):383-386. DOI: 10.1016/j.jcat.2007.02.019

[155] Azizi Y, Petit C, Pitchon V. Formation of polymer-grade ethylene by selective hydrogenation of acetylene over Au/CeO_2 catalyst. Journal of Catalysis. 2008;**256**(2):338-344. DOI: 10.1016/j.jcat.2008.04.003

[156] Capdevila-Cortada M, Vilé G, Teschner D, Pérez-Ramírez J, López N. Reactivity descriptors for ceria in catalysis. Applied Catalysis B: Environmental. 2016;**197**:299-312. DOI: 10.1016/j.apcatb.2016.02.035

[157] Carrasco J, Vilé G, Fernández-Torre D, Pérez R, Pérez-Ramírez J, Ganduglia-Pirovano MV. Molecular-level understanding of CeO_2 as a catalyst for partial alkyne hydrogenation. The Journal of Physical Chemistry C. 2014; **118**(10):5352-5360. DOI: 10.1021/jp410478c

[158] Zhu HZ, Lu YM, Fan FJ, Yu SH. Selective hydrogenation of nitroaromatics by ceria nanorods. Nanoscale. 2013;**5**(16):7219-7223. DOI: 10.1039/C3NR02662K

[159] Chettibi S, Wojcieszak R, Boudjennad EH, Belloni J, Bettahar MM, Keghouche N. Ni-Ce intermetallic

phases in CeO_2-supported nickel catalysts synthesized by γ-radiolysis. Catalysis Today. 2006;**113**(3–4):157-165. DOI: 10.1016/j.cattod.2005.11.058

[160] Kim M, Park J, Kannapu HPR, Suh YW. Cross-aldol condensation of acetone and n-butanol into aliphatic ketones over supported Cu catalysts on ceria-zirconia. Catalysts. 2017;**7**(9):249. DOI: 10.3390/catal7090249

[161] Gürbüz EI, Kunkes EL, Dumesic JA. Integration of C–C coupling reactions of biomass-derived oxygenates to fuel-grade compounds. Applied Catalysis B: Environmental. 2010;**94** (1–2):134-141. DOI: 10.1016/j.apcatb.2009.11.001

[162] Reddy BM, Sreekanth PM, Lakshmanan P, Khan A. Synthesis, characterization and activity study of $SO_2-/Ce_xZr_{1-x}O_2$ solid superacid catalyst. Journal of Molecular Catalysis A: Chemical. 2006;**244**(1–2):1-7. DOI: 10.1016/j.molcata.2005.08.054

[163] Amoroso F, Colussi S, Del Zotto A, Llorca J, Trovarelli A. An efficient and reusable catalyst based on Pd/CeO_2 for the room temperature aerobic Suzuki–Miyaura reaction in water/ethanol. Journal of Molecular Catalysis A: Chemical. 2010;**315**(2):197-204. DOI: 10.1016/j.molcata.2009.09.012

[164] Postole G, Chowdhury B, Karmakar B, Pinki K, Banerji J, Auroux A. Knoevenagel condensation reaction over acid–base bifunctional nanocrystalline $Ce_xZr_{1-x}O_2$ solid solutions. Journal of Catalysis. 2010; **269**(1):110-121. DOI: 10.1016/j.jcat.2009.10.022

[165] González-Arellano C, Abad A, Corma A, García H, Iglesias M, Sanchez F. Catalysis by gold (I) and gold (III): A parallelism between homo-and heterogeneous catalysts for copper-free Sonogashira cross-coupling reactions. Angewandte Chemie International Edition. 2007;**46**(9):1536-1538. DOI: 10.1002/anie.200604746

[166] Sabitha G, Reddy KB, Yadav JS, Shailaja D, Sivudu KS. Ceria/vinylpyridine polymer nanocomposite: An ecofriendly catalyst for the synthesis of 3, 4-dihydropyrimidin-2 (1H)-ones. Tetrahedron Letters. 2005;**46**(47): 8221-8224. DOI: 10.1016/j.tetlet.2005.09.100

[167] Sabitha G, Reddy NM, Prasad MN, Yadav JS, Sivudu KS, Shailaja D. Efficient synthesis of bis (indolyl) methanes using nano ceria supported on vinyl pyridine polymer at ambient temperature. Letters in Organic Chemistry. 2008;**5**(4):300-303. DOI: 10.1016/j.ejmech.2009.09.045

[168] Perlmutter P. Conjugate Addition Reactions in Organic Synthesis. Vol. 9. Oxford: Pergamon Press, Elsevier; 2013. DOI: 10.1002/ange.19931051144

[169] Safaei-Ghomi J, Asgari-Keirabadi M, Khojastehbakht-Koopaei B, Shahbazi-Alavi H. Multicomponent synthesis of C-tethered bispyrazol-5-ols using CeO_2 nanoparticles as an efficient and green catalyst. Research on Chemical Intermediates. 2016;**42**(2): 827-837. DOI: 10.1007/s11164-015-2057-7